HONG KON

THE EXPORTER'S
DISTRIBUTION PRIMER

JAMES KEENAN

Canada	Groupe
Communication	Communication
Group	Canada
Publishing	Édition

CANADIAN CATALOGUING IN PUBLICATION DATA

Keenan, James

Hong Kong/South China: the Exporter's Distribution Primer

Includes bibliographical references.
ISBN 0-660-14925-7
DSS cat. no. K49-2/1993E

1. Hong Kong — Commerce — Handbooks, guides, etc. 2. Hong Kong — Economic conditions — Handbooks, guides, etc. 3. Hong Kong — Social conditions — Handbooks, guides, etc. 4. Canada — Commerce — Handbooks, guides, etc. I. Asia Pacific Foundation of Canada. II. Title. III. Title: The Exporter's Distribution Primer.

HC470.3K43 1993   330.95125′05   C93-099429-9

© Minister of Supply and Services Canada 1993

Available in Canada through

your local bookseller

or by mail from
Canada Communication Group — Publishing

Ottawa, Canada   K1A 0S9

Catalogue No. K49-2/1993E
ISBN 0-660-14925-7

We would especially like to thank

ΞⅡ *ERNST & YOUNG*

for providing guidance and introductions to
their contacts, and for authoring the chapter
"Taxation in Hong Kong".

STIKEMAN, ELLIOTT
Canada's Global Law Firm

for giving us access to their extensive
network in Hong Kong.

Cover photo: courtesy of HKTA

TABLE OF CONTENTS

DREAM OF THE RED CHAMBER

PREPARING THE GROUNDWORK

DISTRIBUTION AGREEMENTS

MANAGING YOUR DISTRIBUTION RELATIONSHIPS

PLANNING AND REFERENCE

ANNEXES

GENERAL INTRODUCTION

Hong Kong/South China: The Exporter's Distribution Primer is a business guide designed for exporters of consumer products or capital goods who wish to know how best to select and manage intermediaries – dealers, distributors, agents and middlemen of all kinds – in order to maximize their chances of success in the Hong Kong and PRC markets. Our belief is that intermediaries play such an important role in the Asia Pacific region that exporters must devote as much time and energy learning how to sell to them as they do trying to understand end-users.

Because this Primer was written primarily for small and medium exporters whose limited budgets allow for few mistakes, we decided to concentrate on Hong Kong distributors active in the PRC as opposed to PRC middlemen. The legal, institutional and behavioral context in the PRC is still too fluid and complex for most small exporters who wish to go beyond hit-and-miss opportunities and achieve some measure of control over their distribution channels.

For the sake of simplicity, the word "he" is used throughout the book to refer to both genders. We hope our female readers will not interpret this as an indication that Hong Kong business life is closed to foreign businesswomen. Indeed, Hong Kong and the PRC are very open in this respect and a large number of local businesswomen are involved in distribution, especially in the area of consumer products.

USING THIS GUIDE

The Exporter's Distribution Primer is written in such a way that it can be read from beginning to end or referred to quickly for facts on topics of special interest. If in doubt about where to look for specific information, consult the Table of Contents or the Index.

ITS LIMITATIONS

Guidebook information is inevitably prone to become outdated. The material contained in this Primer was assembled up to September 1, 1992, and unless otherwise indicated, is based on information available at that time. While the greatest care has been taken in the preparation of this Primer, the Asia Pacific Foundation of Canada, Canada Communication Group, Ernst & Young, Stevenson, Wong & Co. and Stikeman Elliott (hereafter referred to as "the participants") cannot accept any liability for any consequences arising from the use of the information contained herein.

Where an opinion is expressed, it is that of the author and does not necessarily coincide with the views of the participants.

We welcome corrections and suggestions from our readers; please write to:

The Exporter's Distribution Primer
Asia Pacific Foundation of Canada
8160 Sartre St.
Brossard, Quebec
Canada J4X 1P2
Tel: (514) 466-1156
Fax: (514) 466-1156

ACKNOWLEDGEMENTS

So little has been written about distribution that a research project on this subject is doomed to failure unless one is introduced to the right people – suppliers, distributors and consultants – all willing to take time off from their busy schedules to share their experiences. I wish to take this opportunity therefore to thank the many people who helped steer me in the right direction: Michael Adams (Ernst & Young); Heather Allan (Canadian Chamber of Commerce in Hong Kong); Steve Clark (Inchcape Pacific Ltd.); Denis Crépault (APFC-Taipei); Monica Drake (APFC); Steven Gawreletz (Commission for Canada); Brian Hansen (Stikeman Elliott); Mireille Lafleur (Délégation du Québec); Ms. Shelley Lau (Omelco); Arthur McInnis (Faculty of Law, University of Hong Kong); Gary Nachshen (Stikeman Elliott); Arnie Rusinek (Proact International); Ephrem Shiu (CIBC); Arlyle Waring (ConsultAsia); and Sandra Wilking (APFC).

To the many people who agreed to be interviewed, I extend my heartfelt thanks. Their willingness to share their time and expertise with me made my research work a pleasure and a privilege. In particular, I wish to acknowledge the support of a core group of interviewees whose insights were instrumental in guiding my work: Tony Cheung (Anhong Industrial Ltd.); Gilbert Chung (Gemex Trading Ltd.); John Crawford (Ernst & Young); Enzo Cunico (Siber Hegner Marketing); Wilson Law (HKTDC); Lawrence Leung (Canadian Embassy); Simon Ling (BTR Dunlop China Ltd.); Jimmy Mak (Lamko Group); Ms. Juling Ngai (Survey Research Hong Kong Ltd.); Edward Rubin (Corton Hill Investments Ltd.); Louis Tong (SRG China), and Timothy Wu (JDH Trading Ltd.).

I am similarly grateful to Ramon Archer (JDH); George Baeder (Pac Rim); David Bottomley (Asia Commercial Research); David Brodess (Business International); Harold Chan (Hagemeyer); Ginette Charbonneau (APFC-Quebec); Debora Chatwin (Artistree Hong Kong Ltd.); Henry Chong (The Peking Craft Company); Michael Chow (JDH); Barry Cousins (Inchcape Pacific Ltd.); Frank Davies (SRG China); Philip Day (Pac Rim); Lionel Desjardins (DMT Securities Co.Ltd.); Peter Dove (Park 'N Shop); Josef Fieg (Wellcome); Neil Fifer (Gilman Office Machines); Nicholas Fincher (Frank Small & Associates); Katherine Forestier (South China Morning Post); Lam Fu (Lamko Group); Stuart Hampton (ASM Group); John Henderson (Pacific Rim); Helmuth Hennig (Jebson & Co.Ltd.); Jimmy Ho (JDH); Terry Hum (Ernst & Young); Wendy Ip Lui Wan (Schmidt & Co.); Kathleen Jones (Asian Commercial Research); Avis Kong (Marketing Decision Research); James Ku (Inchcape Pacific Ltd.); Lucy Kwan (Huang Kuan & Associates Ltd.); Anna Lai (HKTDC); Alain Larocque (Technomic Consultants International); B.Y. Lau (Edward Keller Ltd.); Dr. Clint Laurent (Asia Market Intelligence Ltd.); Chun-Ming Lee (Claridge House Ltd.); Gabriel Lee (Jardine Consumer Products); Kenrick Leung (International Research Associates); Bannie Lee (Arnhold & Co.Ltd.); Kwan Li (Ernst & Young); Michael Liu (Stevenson, Wong & Co.); H.Y. Lo (Goodman

Medical Supplies Ltd.); Saul Lockhart (HKTDC); Charles Longley Jr. (Getz Bros. & Co.); Kenneth Luciani (EAC Consumer Products); Helmut Luehrs (Jebson & Co. Ltd.); Barry Macdonald (Coopers & Lybrand); George Mak (Nu Skin Hong Kong Inc.); Billy Man (Jardine China Consumers Products); Bernard Pouliot (DMT Securities); Roger Pyatt (University of Hong Kong Business School); Steven Rasin (Technomic Consultants International); Peter Rhodes (Faculty of Law, University of Hong Kong); Chris Robinson (MBL Asia-Pacific Ltd.); Mike Rushworth (Crown Motors Ltd.); Peter Scholz (Jacobson van den Berg, Far East, Ltd.); Lily Shum (HKTA – Canada); Maisie Shun Wah (The Swire Group); Tommie Sutanu (Odyssey International (Trading) Ltd.); Leslie Tam (Leslie Tam & Associates); Tan Wanhong (Sime Darby Hong Kong Ltd.); Shirley Tang (DATEX Pacific Ltd.); Anthony Troughton (Ernst & Young); Arthur Tsoi (Dah Chong Hong Ltd.); Chris Walshe (Wiggins Teape (Hong Kong) Ltd.); Anne Whetham (Canadian Embassy); M. Williams (Hongkong Bank); Allen Wong (Swire Systems Ltd.); Brian Wong (Commission for Canada); Daisy Wong (Harry Wicking); Dickson Wong (Ernst & Young); Marina Wong (Nu Skin Hong Kong Inc.); Patric Wong (Jardine Consumer Products); W.H. Wong (HKTDC); Andrew Wu (Ernest & Young); Annie Wu (World Trade Centre Club Hong Kong); Victor Yang (Boughton Peterson Yang Anderson); Zita Yau (Commission for Canada); Stephen Yee Bit Yan (JDH); Bernard Yeung (Commission for Canada); William Yip (Canada Land Ltd.); and Zhang Xiao-ling (Radio Canada International).

Last, but not least, I wish to thank Kerry Bingham from the New Brunswick government's Trade and Investment Division whose many comments on the first draft of this book were very helpful, and Shirley Zussman, the editor of this book, whose unfailing professionalism made up for the author's exasperation with his own text.

GLOSSARY

ADB	Asian Development Bank.
Agent	A person or firm that has been granted the right to bind a supplier contractually in certain carefully-defined sales transactions. See the section "Defining Relationships" for more details.
Allowance	An amount or percentage given to a distributor or retailer to reduce a payment due to stock shrinkage, damage, breakage, spoilage, impurities, ·etc.
AmCham	The American Chamber of Commerce in Hong Kong.
APFC	The Asia Pacific Foundation of Canada.
AWSJ	Asian Wall Street Journal.
Basic Law	The "mini-constitution" which will govern Hong Kong's administration after 1997. It was drafted by a joint HK-PRC committee and accepted in March 1990 by the National People's Congress in Beijing.
BI	Business International.
BOC	Bank of China which once was the PRC's only foreign exchange bank. It should not be confused with the People's Bank of China, the PRC's central bank.
Canton	See Guangzhou.
Catalogue price	See list price.
CBR	The China Business Review (US-China Business Council). See Annex 7.
CCP	Chinese Communist Party.
CETRA	China External Trade Development Council, the equivalent of HKTDC in Taiwan.
Chong Hwa Travel Agency	Taiwan's *de facto* representative office in Hong Kong.

CIF	Cost, insurance, freight. "CIF Hong Kong" means that the seller's price includes all charges and risks until the ship carrying the goods arrives in Hong Kong. From that point, the buyer must bear all charges and risks, including unloading costs, lighterage and wharfage, unless the sales contract specifies otherwise. With Hong Kong distributors, sales quotations are usually given either CIF or CIF port of shipment, plus actual charges for freight and insurance.
CITIC	China International Trust and Investment Corporation (PRC).
CJV	Contractual joint venture. The overwhelming majority of CJVs in the PRC are formed with Hong Kong firms and are located in Guangdong province. Unlike equity joint ventures, CJVs are similar to partnerships where no separate "legal person" is formed.
Compatible products	Products which do not compete with each other but are intended for the same type of customer, e.g., car engines and spark plugs.
Compensation trade	The supply of equipment or technology to a PRC enterprise by a foreign firm in exchange for goods produced with that equipment (direct compensation) or other goods (indirect compensation or countertrade).
Consignment	The act of entrusting goods to a dealer for sale while retaining ownership of them until sold. The dealer pays only when the goods are sold.
Cooperative venture	A business entity which combines elements of compensation trade and JVs in projects where returns are more long-term. The PRC partner typically supplies factory premises, labor and raw materials while the foreign partner contributes equipment and capital. The Chinese manage the project and repay the foreign partner with products or cash.
CTR	China Trade Report (see Annex 7).

D&B	Dun & Bradstreet (HK) Ltd.
Discount	An amount deducted from a payment that is due when the buyer pays in cash (cash discount), makes early payments (trade discount), buys in quantity (quantity or volume discount), or grants some other advantage to the seller.
Distributor	An independent company which takes ownership of its supplier's merchandise, maintains an inventory, and deploys its own sales, marketing and after-sales service staff in passing the goods down the distribution chain. See the section "Defining Relationships" for more details.
Domestic exports	In Hong Kong trade statistics, exported goods which are either naturally produced in Hong Kong or the result of a manufacturing process which has permanently changed the shape, nature, form or utility of imported raw materials. Processes such as diluting, packing, bottling, drying, assembly, sorting, decorating, etc., do not transform imported raw materials sufficiently to qualify as Hong Kong origin.
EAC	The East Asiatic Company (Hong Kong) Limited.
EJV	Equity joint venture. A limited liability company established by two or more partners that pools assets to create a separate legal entity for the purpose of undertaking a specific business. Hong Kong investors have generally avoided this kind of investment mechanism in the PRC because contractual joint ventures (see CJV) offer them more control and flexibility.
Entrepot trade	The business of importing goods for the purpose of processing or reprocessing and/or repackaging them, exclusively for export. These activities frequently involve consolidation, deconsolidation, and transshipment (see Re-Exports and Transshipments).

EXCO	Executive Council. EXCO plays a role similar to the cabinet in a British parliamentary system. The council normally meets *in camera* once a week, and its proceedings are confidential although many of its decisions are made public. Theoretically, its role is to advise the governor on policy matters; in practice, it functions as a corporate decision-maker. It has 10 members, all of whom are appointed by the governor.
Export processing	This includes both the processing of imported raw materials for export and the assembly of imported components to produce final goods for export. Most Hong Kong CJVs in the Pearl River Delta are export processing operations.
FEC	Foreign exchange certificate. Chinese currency is denominated in yuan and is available in two forms: the Renminbi (Rmb) which is not convertible and the foreign exchange certificate which is convertible. The two currencies are theoretically at par but the FEC trades at a premium on the black market. It is worth noting that the Hong Kong dollar has replaced the FEC in Guangdong as the *de facto* convertible "Chinese" currency of choice.
FEER	The Far Eastern Economic Review (see Annex 7).
FIFO	First-in, first-out. In Hong Kong, the cost of inventory for tax purposes is determined according to the FIFO principle.
FMCP	Fast-moving consumer product.
FOB	Free on board. Transportation term which means that the invoice price includes delivery at the seller's expense and risk up to a specified point (often a port) and no further. Title normally passes from seller to buyer at the FOB point by way of a bill of lading.
FOREX	Foreign exchange.
GATT	General Agreement on Tariffs and Trade.

GDP	Gross domestic product. The total value of a country's annual output of goods and services.
GNP	Gross national product. GDP plus income derived from property held abroad as well as the economic activity of residents abroad, minus the corresponding income of non-residents in the country.
Godown	A warehouse.
Gross (profit) margin	1. An amount. The dollar difference between sales (called turnover in HK) and the cost of goods sold during a stated period of time. Also called gross profit. 2. A percentage. Gross profit as a percentage of sales. Also referred to as gross margin percentage.
GSD	The Government Supplies Department (HK). Purchasing is done by its Procurement Division.
Guangdong	The Chinese province immediately adjacent to Hong Kong with a population of approximately 64 million. Its capital is Guangzhou (Canton).
Guangzhou	Canton. The capital of Guangdong Province.
HK	Hong Kong. HK$50 means 50 Hong Kong dollars.
HKETO	Hong Kong Economic and Trade Office.
HKTA	Hong Kong Tourist Association.
HKTDC	Hong Kong Trade Development Council. Founded in 1966, it is responsible for expanding the markets for Hong Kong's manufactured goods, promoting Hong Kong's products, and creating a favorable image for Hong Kong as a trading partner and a manufacturing center. HKTDC has a network of offices around the world (see Annex 2).
hong	Large trading house.

HS	Harmonized Commodity Description and Coding System, or more simply, Harmonized System. As of January 1, 1992, Hong Kong's official trade statistics are classified according to the HS product coding system. The system formerly in use was SITC.
I/E corporation	Import/export corporation (PRC). There were 9 national I/E corporations in the late 1970s. Today, there are over 5,000 tied to every level from the central government to single enterprises.
II	Illegal immigrant.
Imports	In Hong Kong trade statistics, all goods imported into Hong Kong for domestic consumption or for subsequent re-export.
Indenting	Selling on behalf of a supplier for a commission without taking title or possession of goods. While the buyer normally pays the supplier directly, Hong Kong practice frequently alters this formula. Goods are sometimes accepted by an intermediary on a short-term consignment basis, with payments made to an intermediary who pays the supplier after taking a commission. Also called "doing indent business" or "selling on an indent basis".
IASC	International Accounting Standards Committee.
IPR	Intellectual property rights.
IRD	Inland Revenue Department (HK).
JDH	Refers to companies within Inchcape Pacific Limited responsible for importing, exporting and retailing within Hong Kong and China. All are headquartered in the JDH Center situated in Shatin.
JV	Joint venture.
KCR	The Kowloon-Canton Railway.
KISS	Keep it simple and straightforward.
KMT	Kuomintang. The Nationalist Party in power in Taiwan.
L/C	Letter of credit.

LEGCO	Legislative Council. Its primary function is the enactment of legislation which becomes law once the governor has given his assent. LEGCO is much larger than EXCO, with 60 members half of whom are appointed by the governor and half elected by 15 functional constituencies. It meets once a week and its meetings are televised.
LIFO	Last-in, first-out. In Hong Kong, LIFO is not used to determine the cost of inventory for tax purposes.
List price	The published or marked price, what the end user is asked to pay. Also called retail price, catalogue price, or suggested selling price.
Listing fee	A one-shot payment Hong Kong retailers sometimes ask of suppliers before giving them access to shelf space. Also called slotting fee.
LRT	Hong Kong's light rail transit.
Manufacturer's representative	A person or firm that arranges or facilitates sales for its suppliers without the right to bind them contractually. Manufacturer's representatives are paid on a commission basis and do not normally take ownership of the goods they sell. For more details, see the section "Defining Relationships".
Markdown	A dollar or percentage reduction in retail price, usually when an item cannot be sold at the list or suggested price due to damage, soiling, style changes, excessively high list prices, etc. Percentage markdowns can sometimes be ambiguous. If the price is reduced from $100 to $75, the markdown may be expressed as a percentage either of the former price ($25/$100, i.e., 25%) or of the reduced price ($25/$75, i.e., 33%).

Markup	The dollar or percentage increase between the buying price and the selling price. In dollar terms, it is identical to the gross profit. In percentage terms, however, it can be computed either the conventional way, i.e., as a percentage of the retail selling price, or as a percentage on cost. A hat bought for $10 and resold for $15 can be said to have a 33% markup on price ($15-$10=$5; $5/$15=.33) or a 50% markup on cost ($5/$10=.50).
MFN	Most favored nation. According to GATT's "trade without discrimination" principle (also known as the "most-favored-nation clause") all parties that are signatories of the GATT are bound to grant each other treatment as favorable as they grant to any other nation in the application of import and export tariffs. Although the PRC is not a GATT signatory, the United States and China granted each other MFN status in 1980. The renewal of this status every year by the US president has become a contentious issue ever since Chinese authorities crushed the democracy movement in 1989.
MOFERT	Ministry of Foreign Economic Relations and Trade (PRC).
MTR	Hong Kong's mass transit railway.
Net price	1. The price actually paid. List price minus all discounts, deductions and allowances. 2. A price on which no discount will be allowed. Net distributor price is the supplier's minimum price for a given territory and should *never* be published in price list form because net prices are highly confidential and vary with different distributors. The only price lists given out by experienced exporters are retail price lists.

Net (profit) margin	1. An amount. The dollar difference between sales and the cost of goods sold during a stated period of time after expenses and taxes have been deducted. Also called net profit. 2. A percentage. Net profit as a percentage of sales (called turnover in HK).
OMELCO	Office of the Members of the Executive and Legislative Councils. It is made up of members of EXCO and LEGCO who, appointed or elected, are not government officials. OMELCO plays an ombudsman role in the administration of Hong Kong. Its members advise on formulation of and change to government policy; consider complaints from members of the public; and monitor the effectiveness of public administration through a number of issue-specific standing panels and special groups.
PADS	The Port and Airport Development Strategy. A series of massive infrastructure projects designed to add anchorage, a new container terminal and a new airport to Hong Kong's present facilities. The financing for the airport portion of PADS has been very controversial in Beijing.
Parallel trade	The international commerce in goods by traders who function outside of the official distribution networks established by the original manufacturers. It is entirely legal. The goods thus traded are referred to as parallel imports (PI) or parallel exports.
PI	Parallel imports. See parallel trade.
PRC	People's Republic of China.
PRD	Pearl River Delta. A belt of 22 coastal open cities around the estuary of the Pearl River, near Hong Kong.
Rebate	Any refund made to the buyer after he has made full payment. Discounts, on the other hand, are made in advance of payment.

Re-exports	In Hong Kong trade statistics, products originally imported into Hong Kong before being exported and have not undergone sufficient transformation while in Hong Kong to qualify them as domestic exports.
Retail price	See list price.
Retained imports	In Hong Kong trade statistics, the difference between total imports and re-exports. This understates the value of retained imports because the re-export margin (the cost of transportation, insurance, storage, packaging and minor processing as well as the profits made by the Hong Kong intermediary) is not counted. The HKTDC has estimated this re-export markup to be approximately 15%. The re-export markup for PRC goods, many being the products of Hong Kong export processing activities in Guangdong, is considerably higher at 25%.
Rmb	Renminbi. Literally "the people's money" (also called yuan). The PRC's currency (see FEC).
ROC	Republic of China (Taiwan).
SAEC	State Administration of Exchange Control. The government body responsible for supervising the PRC's swap centers.
SAR	Special administrative region. According to the Sino-British Joint Declaration (1984), this will be Hong Kong's administrative status within the PRC after 1997.
SEZ	Special economic zone.
SITC	Standard International Trade Classification. Now superseded by HS, SITC was the product classifications system underlying Hong Kong's trade statistics. Trade enquiries to HKTDC using SITC coding are still answered.
Slotting fee	See listing fee.
SSAP	Statement of Standard Accounting Practice (HK).
Suggested selling price	See list price.

Transshipments	Goods which are consigned directly from the exporting country to a buyer in the importing country while being transported through or stored in Hong Kong. Since they are in transit and do not clear customs, transshipped goods are not included in Hong Kong's trade statistics.
Turnover	1. In general, the number of times a stock of merchandise, raw materials, capital, or some other asset is used within a period of time.
	2. Sales, in British and Hong Kong terminology. A firm with annual sales of HK$10 million is said to have a turnover of that amount.
UK	United Kingdom.
Xinhua News Agency	The PRC's *de facto* representative office in Hong Kong.
Yuan	China's currency. It is available in two forms: Renminbi (Rmb) and foreign exchange certificates (FEC).

THE
SOUTHERN GATE

HISTORICAL OVERVIEW

Though nominally free from Chinese influence, Hong Kong's astonishing rise to prosperity during the last 150 years is primarily due to its ability to take advantage of China's changing role in the world economy. However, its political stability, also a key factor behind its success, is increasingly strained by China's aggressive involvement in local administration.

BEFORE 1945

A colony is born

The present territory of Hong Kong was acquired by Great Britain from China in three stages: Hong Kong Island in 1842 after the First Opium War, the Kowloon Peninsula twenty years later as a result of the Second Opium War, and, in 1898, the New Territories (92% of Hong Kong's total land area, comprising a mainland area, Deep Bay, Mirs Bay, and some 236 adjacent islands) by a 99-year lease.

Minimal government

The major reason for Great Britain's interest in Hong Kong was its magnificent harbor, her new colony serving primarily as an important trade link between China and the West. Government administration was minimal and non-interventionist, limiting itself to enforcing law and order and the provision of basic sanitary services. Following the usual Crown colony pattern, political power was in the hands of a governor advised by executive and legislative councils staffed by nominated colonial civil servants. In time, Hong Kong's expatriate and Chinese business elites were granted more influence on local political affairs through selective nominations to these councils; however, they were not allowed to constitute a majority nor directly participate in policy decisions. British colonial policy could best be described as conditional *laissez-faire* with stability as its overriding concern.

A foot-loose population

From 1900 to 1940, when China was ravaged by warlords and torn by Communist insurrection, Hong Kong remained a peaceful entrepot trading center. Because people could move unimpeded across its northern border, the flow of migrants increased whenever there was too much unrest in China; the process was reversed when peaceful conditions were restored. Many migrants also came in search of work and then returned later to their villages. Europeans were similarly transient; few considered Hong Kong their permanent home.

In 1941, however, the Japanese army invaded Hong Kong, imprisoned the British authorities, and virtually destroyed the local economy.

The Territory of Hong Kong

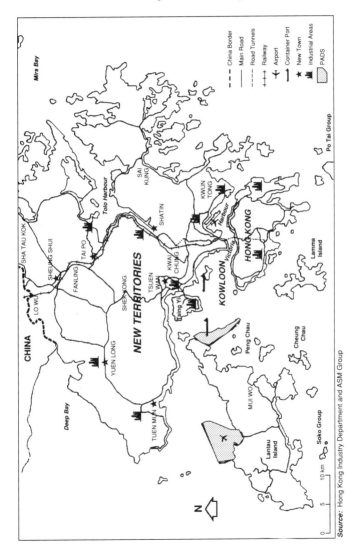

Source: Hong Kong Industry Department and ASM Group

During the war, the colony's population, which had reached over 1 million by 1939, was reduced to 600,000 as a result of privation and mass deportation by the Japanese. British rule was only restored in 1945.

AFTER 1945

Two external shocks

During the next few years, Hong Kong was shaken by two international events which profoundly influenced its economic development. Its population base, heavily depleted during the Japanese occupation, was dramatically boosted following the victory of the Communists in 1949. Over 1 million refugees fled to Hong Kong, thus providing a pool of disciplined, hard-working labor for local industry during the 1950s. An important subgroup among them was exiled Shanghainese textile owners whose capital and organizational skills played a crucial role in Hong Kong's subsequent industrialization.

The second shock to Hong Kong's economy was the Korean War in 1950. The ensuing embargo on trade with China gravely damaged entrepot trade and forced Hong Kong's business community (only 10% of the colony's exports were produced locally) to manufacture for export. The small size of the domestic market gave them no other option.

Migrant pressures

The pressures resulting from the massive influx of refugees from the civil war in China in the late 1940s forced the colony to abandon its policy of free access, and its border was closed in 1950. Since then, Chinese authorities have used population pressures to communicate their displeasure whenever Hong Kong's policies have not been satisfactory, sometimes allowing massive surges in illegal immigration (200,000 people were allowed through in 1979-80) and refusing to accept returned illegals. To this day, China maintains the pressure by allowing 75 legal migrants to cross the border daily, in spite of Hong Kong's representations.

The massive strain on all public services arising from new refugees forced the Hong Kong government in 1953 to construct estates to house them. The housing program has continued since then, especially in satellite towns situated in the New Territories. The government is now building more than 50,000 public housing units per year, of which approximately one-third are for sale. By 1991, more than 50% of the population were living in government-provided housing.

Since 1976, many refugees from Vietnam have attempted to reach Hong Kong by sea. At first, most were accepted for resettlement in North America, Australia and Europe; however, when immigration quotas in those countries were reduced in 1982, Hong Kong was forced to confine all newly-arrived Vietnamese to closed camps. There are now more than 52,000 refugees accommodated in permanent camps. In October 1991, Vietnam and the UK reached an agreement on a mandatory repatriation program.

Democracy shelved

Following the restoration of civil government on the traditional colonial pattern in 1946, the returning Governor promised constitutional reforms that would have provided a greater measure of local self-government. These plans were postponed following the Communist victory in China in 1949, and were finally abandoned in 1952. Until 1984, the British government's cautious approach combined with the lack of interest in politics of the people of Hong Kong to ensure no further moves towards democratic government, largely in deference to China's opposition to any such changes.

Pearl River Delta Economic Zone

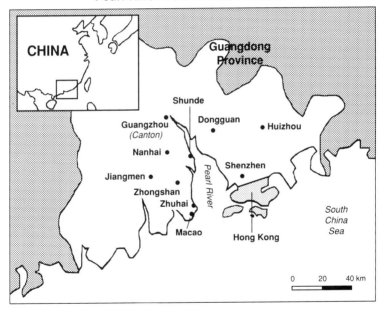

Structural shift

The manufacturing industry which evolved as a consequence of the embargo on trade with China has been dominated, in descending order of importance, by textiles and garments, electrical appliances, plastics and, more recently, watches and clocks. Beginning in the mid-1980s, however, manufacturing lost its dominance to such service industries as tourism, insurance, transportation, shipping and banking. Today, with its service sector accounting for nearly

27

84% of total domestic input, Hong Kong is perhaps the most service-oriented economy in the world. One important reason for this structural shift has been growing US protectionism. Because Hong Kong's manufacturers produce a limited number of end products which are highly dependent on the American market, expansion has been slowed by a growing list of tariffs and quota restrictions.

A much more significant reason for the relative decline of manufacturing has been Hong Kong's progressive reintegration into China's economy in the wake of its trade-oriented reform program (see Reforms under Deng). The result has been a rebirth of entrepot trade as high salaries and surging land prices force manufacturers of non-quota products to relocate their operations into neighboring Guangdong Province, especially in the Pearl River Delta, a belt of 22 coastal open cities situated around the nearby estuary of the Pearl River (see map, p. 27). We shall examine this phenomenon in greater detail in the context of the growing interdependence between Hong Kong, the PRC and Taiwan (see The Birth of Greater China).

POLITICAL DEVELOPMENTS

The Sino-British Joint Declaration

On December 19, 1984, the UK and China signed a joint declaration whereby the whole territory reverts to Chinese sovereignty in 1997. Given Hong Kong's dependence on China

for such basic resources as water and the fact that the lease on the New Territories (92% of Hong Kong's land area) expires in 1997, the UK had little choice but to ensure that the transition be as painless as possible. In this declaration, Chinese leaders pledged that Hong Kong will, under the guiding principle of "one country, two systems" become a Special Administrative Region (SAR) of China and be allowed to carry on its capitalist economic and social system for 50 years beyond 1997. Since then, a Basic Law Drafting Committee composed of representatives from both China and Hong Kong, has drawn up a "constitution" for the future SAR called the Basic Law. After lengthy consultation and redrafting, this law was accepted in March 1990 by the National People's Congress in Beijing.

The Tiananmen stampede

Local confidence in Hong Kong's future, which had followed the signing of the 1984 declaration, was shattered by the massacre of students in Beijing in June 1989 and by the subsequent suppression of the pro-democracy movement throughout China. Demonstrations involving up to one million people were held to protest the slaughter, and emigration overseas jumped from 45,000 in 1989 to 62,000 in 1990.

To stem the emigration stampede and restore confidence in Hong Kong's future, the British government adopted two measures which ultimately backfired. The first was the

British Nationality Scheme which offered full British passports with the right of abode in the UK to 50,000 carefully-selected residents and their families (225,000 individuals). It was hoped that this would encourage key persons to remain in Hong Kong because they had a guaranteed way out if conditions deteriorated. Beijing, however, immediately denounced this move, stating that it would not recognize those passports after 1997. Since these British passport-holders fear discrimination and have no foreign status as far as China is concerned, it is likely that most will leave before 1997.

The airport stalemate

Britain's second move to boost confidence was to announce in October 1989 plans for a series of massive infrastructure projects called PADS (the Port and Airport Development Strategy), designed to strengthen Hong Kong's place as the Far East's premier site for business, trade and tourism. The 10 core projects were billed at HK$98.6 billion in March 1991 and were to be paid from the Hong Kong government's reserves and from a loan to be repaid after 1997. Beijing insisted that its endorsement of the project and the loan would depend on Britain's willingness to consult China on all important issues that straddle 1997, in effect giving China veto powers over the Hong Kong government's activities before 1997 regardless of whether they had any relation to PADS.

Britain and China subsequently reached agreement on a Memorandum of Understanding permitting PADS to proceed as long as the UK guaranteed that a minimum of HK$25 billion would remain in Hong Kong's reserves in 1997 and that government borrowing straddling 1997 would not exceed HK$5 billion. This agreement has since collapsed due to new cost estimates for PADS of HK$112.2 billion amid warnings that it may rise to HK$160 billion by 1997. Beijing also has objections to Britain's strong verbal support of various Hong Kong groups seeking greater democracy. Regardless of PADS' merits, this confidence-building exercise by the UK has dramatically increased Beijing's capacity and willingness to assert its point of view on local affairs. This can only compromise the authority of the Hong Kong government during its last years before 1997.

THE BIRTH OF GREATER CHINA

THE CHINA BUSINESS REVIEW

Growing economic links between China, Hong Kong and Taiwan may eventually reshape the complex economic and political balance of East Asia. This article was originally printed in the May-June 1992 issue of The China Business Review, a bimonthly magazine published by the US-China Business Council (see Annex 7). Its author is Pamela Baldinger, Editor-in-Chief of The China Business Review.

For the past four decades, business people dealing with East Asia have learned to work with three distinct Chinese markets: the vast, central planned economy of the People's Republic, the capitalist enclave on the free-trading British colony of Hong Kong, and the Japanese inspired,

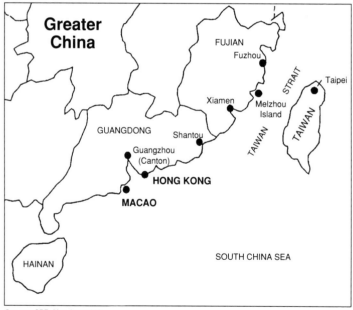

Source: CBR, May-June 1992.

newly industrialized economy of Taiwan. The three areas had relatively little commercial contact with each other until China re-opened its doors to foreign trade and investment in 1978: by the mid 1980s, Hong Kong had firmly established itself as a bridge between its Communist neighbor and Western trading partners. The Nationalist government on Taiwan, in contrast, held steadfast to its policy that Taiwan citizens should have no contact with the enemy regime in China.

By the late 1980s, however, a combination of political and economic considerations – primarily the rising cost of labor and manufacturing in Taiwan – led Taipei to change its tune. Now, the "state of war" across the Taiwan Strait is officially over, indirect trade is booming, and Taiwan is one of the largest investors in its former enemy. The island has also stepped up its presence in Hong Kong, largely in order to deal more effectively with the mainland.

The economic links among China, Hong Kong, and Taiwan are growing at a stunning pace, and promise to eventually weave the three societies into a single economic entity – "Greater China". Already, officials in Taipei and Beijing are struggling to exert political control over the forces of economic integration, with varying success. US businesses and policymakers must also begin to adjust their thinking, as they can now no longer look at any of the three constituent pieces of Greater China without also considering the growing links between them.

Hong Kong and China: rapidly blurring boundaries

Economic relations between China, Hong Kong, and Taiwan are characterized by the transfer of labor-intensive manufacturing operations from Taiwan and Hong Kong to the mainland. The division of labor is well defined: Taiwan and Hong Kong provide technology and management expertise, China supplies cheap, abundant labor, and Hong Kong (and to a lesser extent, Taiwan) provides value-added services such as packaging and transportation.

This combination has led to explosive growth – Hong Kong's total trade leaped from around US$69 billion in 1978 to US$99 billion in 1991, while China's grew from US$21 billion to US$136 billion over the same period – and also to significant restructuring of Hong Kong's economy. According to a 1991 survey of Hong Kong firms conducted by the Hong Kong Trade Development Council (HKTDC), the number of manufacturing establishments in Hong Kong, as well as the number of people employed in such concerns, dropped by 25 percent from 1980 to 1991. During the same period, however, the number of trading entities grew by almost five times to nearly 70,000 – primarily to handle the surge in China's exports. The activities of Hong Kong trading companies extend far beyond the traditional entrepot functions of arranging shipping and insurance; they also include manufacturing-related services such as quality control, packaging, product design, and sample-making.

These activities support the production of thousands of Hong Kong investments and outward processing arrangements on the mainland. There are now some 25,000 enterprises in Guangdong – employing 3 million workers – producing goods for Hong Kong companies. About two-thirds of these workers are engaged in outward processing; the rest work for Hong Kong joint ventures or wholly foreign-owned enterprises. About 80 percent of Hong Kong's China investment – which now exceeds US$10 billion – is in Guangdong Province. According to Guangdong officials, about 36 percent of Hong Kong's industry has moved across the border to the Pearl River Delta.

The shift of Hong Kong manufacturing to China is reflected in the pattern of Hong Kong's trade: over the last several years re-exports have grown at a much faster pace than domestic exports (see Fig. 1-1). Re-exports now account for about 70 percent of Hong Kong's total exports, compared to 45 percent in 1986. Of the total export value of the nearly 3,000 companies surveyed by the HKTDC, Chinese-origin products accounted for 58 percent, compared to the 22 percent for Hong Kong-produced products. The ratio was equal in the HKTDC's 1988 survey.

Investment between Hong Kong and China has not been one-sided. China has invested over US$12 billion in Hong Kong real estate and other sectors, making it the largest outside investor in the territory. Beijing has also been playing a much more vocal role in the political life of the territory, openly criticizing government spending schemes and various democratic features of the colony. In so doing, Beijing is clearly trying to exert its influence in Hong Kong prior to the territory's 1997 reversion to Chinese sovereignty. While this has caused alarm in some quarters, it has not slowed the pace of economic integration. In fact, Beijing's actions seem calculated to drive home a message that has not been missed by the business community – that the future of Hong Kong lies to the north, not the west.

While Beijing is not yet responsible for Hong Kong's foreign relations, the economic integration of Hong Kong and South China means that Hong Kong is increasingly affected by China's political and economic relations with other countries. If the United States – South China's largest export market – were to revoke China's Most Favored Nation status, for example, the Hong Kong government estimates that the territory would lose up to US$15.7 billion in overall trade, up to US$2 billion in income, and as many as 60,000 jobs. Re-exports from China to the United States would probably fall 35-47 percent (US$4.6-6.2 billion). Hong Kong government and business groups have therefore sent numerous delegations to the United States of late to educate US policymakers on the implications of their actions.

Figure 1-1

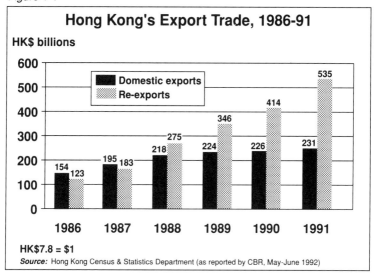

Hong Kong's Export Trade, 1986-91

HK$ billions

HK$7.8 = $1

Source: Hong Kong Census & Statistics Department (as reported by CBR, May-June 1992)

Taiwan and China: breaking the ice

While Hong Kong has clearly been the primary mover of capital light industrial technology to China over the past decade, Taiwan has become a major player in the last few years, thanks to political and economic liberalization at home and a red carpet from China. In 1991, China-Taiwan trade conducted, through Hong Kong – there are no direct shipping or air links between the two – hit US$5.8 billion, up 43 percent from 1990. Taiwan's imports increased 47 percent to US$1.1 billion, while its exports rose

42 percent to US$4.7 billion. The real bilateral trade figure is undoubtedly higher, as a significant amount of business is conducted on the sly or passes through other areas such as Singapore. Taiwan's surplus with the mainland is generated by exports of light industrial goods such as manmade filament, yarn, machinery, electrical and electronic parts, and plastic materials. Its imports generally consist of natural resources and raw materials.

Analysts predict that Taiwan-China trade will hit US$7-8 billion in 1992 which should still place it well below the ceiling of 10 percent of total trade informally imposed by Taipei. The Nationalist government fears anything above this figure would result in excessive dependence on the mainland market and give Beijing undue leverage in its dealings with the island. Under pressure from business interests, however, Taipei has continuously increased the number of goods it permits to be legally imported from China, and the list now totals some 250 items, up from 50 in 1988. Most are raw materials, though some semifinished products were added in late 1991. Despite the increase in the number of products that may be imported legally, nearly a quarter of Taiwan's imports are thought to consist of goods not on the list.

For instance, oil trade between China and Taiwan is thriving, though its existence is officially denied, Chinese gasoline is typically shipped to Singapore, where it is refined and shipped to Taiwan. Each month Taiwan imports four to six shiploads of 200,000 barrels each. Fuel oil from Sinochem's processing facility on Okinawa is supposedly also being shipped to Taiwan by a Taiwanese trading company.

This pattern, whereby trade (and investment) precede official recognition and sanction, has been the norm in Taiwan, where the Nationalist government must toe the difficult line of maintaining the island's economic competitiveness and prosperity while not allowing it to become dependent on the mainland market. In most cases, the regime has sought a face-saving way to recognize reality without appearing too liberal on issues regarding commercial links with the mainland; in February, for instance, Taiwan's high court recognized the Chinese yuan as legal tender, on the grounds that it was circulated and used by citizens in territory claimed by Taiwan. In other examples, such as the decision in December 1991 to let Taiwan exporters negotiate letters of credit with local banks for mainland trade, the motivation seems to be to try to maintain some semblance of control over contact with the mainland, as well as to ensure that Taiwan maximizes its profit from the relationship.

For the most part, only privately held companies have received permission to openly trade with the mainland; now, however, it appears that State-owned companies, which want equal access to cheap Chinese imports to stay competitive, will also gradually gain that right. Taiwan Power, one of the largest publicly held companies in Taiwan, is expected to begin importing coal from China via third parties. The coal probably will have to be purchased on the spot market and must not exceed 30 percent of the company's total spot-market purchases; moreover, the utility's consumption of coal from the mainland must not exceed 20 percent of its total annual purchases. Private companies have been importing Chinese coal since 1988, and in 1991 imported 1.4 million tons (out of 18.6 million total), according to the Energy Commission of the Ministry of Economic Affairs.

No qualms for Taiwan investors

The issues of Taiwan investment in the mainland is more difficult for Taipei to manage. As with trade, the government first turned a blind eye to a phenomenon that was already becoming commonplace, then announced in 1990 that labor-intensive and sunset industries – which skyrocketing labor rates were making non-competitive – would be allowed to invest indirectly in the mainland with official sanction. It then went a step further and announced that Taiwan firms and business delegations investing in the mainland should register with the government. Since then, Taipei has approved investments totaling US$800 million by 28,000 companies; now it is calling for an investment protection treaty with China.

Estimates of total Taiwan investment in the mainland are sketchy, as many of the investments are small and were made secretly. Most China watchers put the figure at around US$2-3 billion, US$1 billion of which was probably committed in 1991. According to Chung Chin, an economist at the Chung Hwa Institution for Economic Research in Taiwan, the figure should be around the same this year.

Around 80 percent of Taiwan's investment in China is in neighboring Fujian Province, where residents speak the same language and often have ancestral links with Taiwan natives. The pattern of Taiwan's investment in China is basically the same as Hong Kong's, though it has

evolved over a much shorter period of time: Taiwanese first tended to lease workshops for processing arrangements, then, increasingly, began to buy the workshops. Now they are more likely to lease land to build their own factories.

Most of the investment has been in light industrial/consumer products, such as footwear and sports equipment. Taipei recently announced it would also approve investment in such service industries as restaurants and entertainment, though such investments have already been taking place for some time. Moreover, Taiwan policymakers have said they will eventually approve investment by banks, insurance companies, and securities firms on the mainland, but no timetable for removing the current ban exists. The Taiwan government has remained steadfast in its refusal to countenance investments by Taiwan companies in strategic or high-technology sectors, though this policy is difficult to enforce and will only grow more so with time.

The mainland has been extremely welcoming of Taiwan investment, with numerous cities and provinces setting up special investment zones for Taiwan investors. Perhaps the most ambitious development along these lines was the recent announcement that a special port for Taiwan trade would be constructed on Fujian's Meizhou Island – the first such port in China. Located north of Xiamen and south of Fuzhou, Meizhou is less than 100 nautical miles from Taiwan and is considered the hometown of Mazu (the goddess of the sea), a popular Taoist deity on Taiwan. Beijing has committed about US$5.5 million to upgrade the port so that it may receive 3,000-5,000 tonne ships; eventually, it will be expanded to accommodate 10,000 tonne ships.

The island is to be developed in two phases by Hong Kong's Huayang Co., with additional funds from Taiwan, Singapore, and Hong Kong investors. The entire project is valued at around US$130 million. The first phase, which is to take place from 1992-96, will feature construction of a 3 km bridge connecting the island to the mainland, development of the Mazu temple area and a golf course, and installation of the basic infrastructure. Phase 2, which is to end in the year 2000, will complete the infrastructure and further develop real estate, tourist facilities, and commercial and service industries.

The growing economic links between Taiwan and China are both a function of, and contributor to, decreasing hostilities between these two traditional enemies. While there have been no political breakthroughs as yet, it appears to be only a matter of time before official contact is made between the two. In the meantime, it will become increasingly difficult for either regime to contain the development of commerce across the Taiwan Strait; competitive pressures will ensure the increasing internationalization of Taiwan companies and liberalization of Taiwan's financial system, thereby eroding Taipei's ability to

rein in Taiwanese investment in the mainland. The mainland, eager for Taiwan investment and technology, will find itself increasingly confronted by the "bourgeois influences" that have accompanied this investment – prostitution, conspicuous consumption, and smuggling. Even more important, perhaps no other group poses quite the same threat of "peaceful evolution" than the Taiwanese.

Hong Kong and Taiwan: closing the triangle

As relations between Taiwan and China have improved, Taiwan has shown more interest in Hong Kong, both as an entrepot and in its own right. Previously, Taiwan commercial relations with Hong Kong were limited, primarily due to Hong Kong's future reversion to Chinese sovereignty. Now, Taiwan is Hong Kong's fourth largest trading partner and its largest source of overseas visitors, accounting for over 20 percent of all arrivals in 1991. Hong Kong has become Taiwan's second largest export market, though 40 percent of the goods are re-exported to China.

Signs of Hong Kong's newfound importance to Taipei are numerous. Perhaps most notable, in September 1991, Taipei sent John Ni, the former director-general of the Industrial Development and Investment Center in the Ministry of Economic Affairs, to Hong Kong as Taiwan's senior representative. Ni was responsible for foreign investment in Taiwan and Taiwan investment abroad, and he clearly will be keeping a close eye on Taiwan's growing presence on the mainland.

The China External Trade Development Council, Taiwan's trade promotion organization, also set up an office in Hong Kong last year, in part to service Taiwan investments throughout East and Southeast Asia, including China. Taiwan's three largest banks also opened offices in Hong Kong in 1991, and the Chinese National Federation of Industries established the Hong Kong-Taipei Business Cooperation Committee with the Hong Kong General Chamber of Commerce. The function of this group is to promote trade, investment, and other economic cooperation between the two business communities. So far, more than 60 Hong Kong companies have expressed interest in joining the committee.

The future of Greater China

Full economic integration of Hong Kong, China, and Taiwan cannot be achieved until some sort of political accommodation is reached between Taiwan and China and restrictions on market access, capital transfer, and communications links between the three areas are lifted. Clearly, not all of these things will happen in the near future, nor will they all happen at once. Progress will be made incrementally, but the momentum of the forces at work and the complementarity of the economies involved should ensure that the process continues.

Already, economists in the three areas are devoting themselves to studying the issues of Greater China, though they don't necessarily agree on how the integration will proceed or how the final entity will function. Disagreement on the respective roles of each of the players was evident at a January 1992 conference on the subject in Hong Kong. Hosted by the Democracy Foundation of Taiwan and the Hong Kong Baptist College, the conference invited economists from all three areas to discuss a framework for economic integration. Despite the discord between the attendants from Taiwan and the mainland, the delegates all agreed that they wanted another congress next year; some suggested establishing a permanent coordinating agency to monitor the process of integration.

Economists, businesspeople, and policymakers outside the region would also be wise to study the ramifications of the emergence of Greater China, for though the integration process might not be particularly smooth or coordinated, its impact will be great. According to *Business Week*, the combined economic output of Taiwan, Hong Kong, and South China (southern Guangdong and Fujian Provinces) was US$275 billion in 1990, and will be on a par with France by the year 2000. Moreover, each of the three areas is already an important trading power in its own right; combined, they undoubtedly would create a new global economic force.

TABLE 1-2 China's trade balance as seen by Beijing...

China-US Trade
(US$ billion)

	1987	1988	1989	1990	1991
Exports to US	3.0	3.4	4.4	5.2	6.2
Imports from US	4.8	6.6	7.9	6.6	8.0
China's trade balance	-1.3	-3.3	-3.5	-1.4	-1.8

Source: China's Customs Statistics (as reported in CBR, May-June 1992)

APFC FOOTNOTE

The emergence of Greater China is introducing many new complexities into US policy towards China, Hong Kong and Taiwan. Three problems are of particular importance to exporters: bilateral accounting, strategic goods and punitive sanctions.

Bilateral accounting

As the United States' trade deficit has soared over the last several years, its balance of trade statistics with individual countries have received increased scrutiny. Analyzing US-China trade can be fairly complicated because so much of it is conducted through Hong Kong. The task is further complicated by the fact that both sides do not account for re-exports in the same way. Chinese statistics (see Table 1-2) consider Chinese goods transshipped through Hong Kong for the US as exports to Hong Kong but takes account of all US imports regardless of whether they come via Hong Kong. The US does the same thing in reverse (see Table 1-3). It considers Chinese goods shipped to the US to be Chinese exports, and American goods transshipped through

TABLE 1-3 ... and Washington

US-China Trade
(US$ billion)

	1987	1988	1989	1990	1991
Imports from China	6.9	8.5	12.0	15.2	19.0
Exports to China	3.5	5.0	5.8	4.8	6.3
China's trade balance	3.4	3.5	6.2	10.4	12.7

Source: Department of Commerce (as reported in CBR, May-June 1992)

Hong Kong for China to be exports to Hong Kong. As a result, there is a wide discrepancy between their respective views of US-China trade and this discrepancy has been a source of trade friction for both sides.

Strategic goods

Inter-regional trade of controlled strategic goods is another phenomenon of increasing concern to US authorities. It is becoming more and more difficult to control the re-sale of strategic technology once it has been sold to Taiwan or especially Hong Kong, regardless of US export restrictions.

Punitive sanctions

The failure of a recent attempt within Congress to punish China for its record on human rights, missile proliferation and trade practices by targeting its state enterprises (as opposed to private enterprises) for sanctions was due in part to the administrative impossibility of disentangling trade relations between China, Hong Kong and Taiwan. The revocation of China's Most Favored Nation (MFN) trading status has the same drawback. Increasingly, punishing China means punishing Taiwan and particularly Hong Kong, two old US allies.

UNDERSTANDING
HONG KONG
CUSTOMERS

THE HONG KONG CONSUMER

Because it is a mature market and a free port, Hong Kong may appear forbidding to foreign exporters of consumer products. However, Hong Kong can be a very receptive market and an ideal testing ground for the rest of Asia as long as exporters compete against PRC imports with innovative designs rather than with lower prices. We wish to thank the Hong Kong market research firm, SRH, which supplied us with most of the statistics and ideas for this section; any errors or omissions are the author's responsibility.

BACKGROUND

The negative side

Hong Kong's economy is clearly at a crossroad. Over the last four years, it has been suffering from a combination of high inflation (7.4%) and low annual GDP growth (4% compared to 7% in the 80s). Moreover, its labor-intensive manufacturing base is moving to Guangdong Province to take advantage of the PRC's lower labor costs and cheaper land.

The malaise caused by these figures has been exacerbated by the far bigger issue of political reintegration into the PRC in 1997. People expect education standards to slide, security to fall, crime rates to rise and freedom to decline as the date for transfer of sovereignty approaches. With the imminent change of employer, public servants in particular are becoming more uneasy and militant.

The positive side

Hong Kong is strategically located with respect to China, a massive economy progressively being reintegrated into the world trading system. It is also at the center of the Asia Pacific region, economically the fastest growing part of the world. Finally, its entrepreneurial environment remains largely intact. Its government is non-interventionist and relatively free of corruption. Taxes are low and the law is respected. Hong Kong also has a very good transportation and communication infrastructure.

DEMOGRAPHIC TRENDS

Cramped

According to the 1991 population census which excludes residents temporarily away (151,833), transients (35,823), and the Vietnamese boat people (51,847), Hong Kong has a population of 5,522,281. This population lives in a mountainous area of 1,070 km^2 (one-third the size of Rhode Island) where the population density can vary from 20,300 per km^2 in Kowloon to 1,920 per km^2 in the New Territories. With smaller families, a trend towards later marriages, an aging population and emigration running at 60,000 a year, any population increase is likely to be negligible during the next five years. Hong

Kong's extremely high population density has obvious consequences for consumer buying habits: except for staples such as rice and cooking oil which are bought in units larger than the North American norm, everything else is bought in smaller sizes because refrigerators and storage cabinets are small in size. The habit of frequent shopping trips to buy small amounts is reinforced by the southern Chinese preference for meals with a large selection of small dishes.

Aging

The median age, which divides Hong Kong's population into equal halves, will advance six points in the 90s, from 31 to 37. In numerical terms, this means that there will be 100,000 fewer children and teenage consumers, and 350,000 more consumers aged 50 years or over by the year 2000.

Population distribution

The slow exodus of the population to the New Territories will continue and accelerate with the building of new roads. The New Territories will be home to half of Hong Kong's population by the year 2000. The demographic division in 1990 was 22% on Hong Kong Island, 43% in Kowloon and New Kowloon, and 35% in the New Territories.

Income

Assuming an inflation rate averaging 8% and real growth of 3%, the average monthly household income will rise from its 1990 level of HK$11,000 to HK$25,500 by the year 2000. These figures must, however, be understood in the context of growing disparities between the wage rates offered to different groups. Between 1982 and 1991, average real wage rates in the manufacturing and trade sectors (60% of total employment) rose by only 10%. In the same period, they rose 35% in transport, business and personal services, and 65% among mid-level managers and professionals. Actual earnings rose faster than wage rates, but that was due to longer working hours. The remarkable gains among managers and professionals probably reflect a shortage caused by emigration at least as much as structural changes in the economy.

Sophisticated

Generally higher educational levels and a smaller discrepancy in education between men and women are having an influence on the average consumer's willingness to try new products. Migration trends and the internationalization of business have also given more Hong Kong people the chance to interact with foreigners. Hong Kong shows one of the world's highest incidences of air travel to other countries, with over one-third of the population having visited another country besides China and Macao.

Replacement market

Hong Kong households, despite their small size, already own an impressive list of appliances. While the market for the more established appliances is one of replacement and upgrading rather than addition, the potential for new innovations remains very good.

CONSUMER VALUES

Money-minded

As 1997 comes closer, Hong Kong people are becoming ever more money-conscious. Everything in life – success, job prospects, gifts, general satisfaction – tends to be reduced to this simple denominator. Political insecurity is leading some to put money aside to finance their emigration. High inflation encourages others to invest in property. Except for the very wealthy, these trends result in less disposable income for frills.

One of the crowd

Perceived popularity is very important for most Hong Kong consumers because they wish to conform rather than stand out. Advertising strategies hoping to target this market stress the sense of confidence and social acceptance which come from using an established brand. In product after product, 70% of the market is occupied by these established brands. Consequently, the battle for survival among new products is usually played out with the 13% of the population more educated and willing to try new things. Failure with this group spells death.

Prestige

Prestige is another key value for Hong Kong consumers because face and appearances are very important to the Chinese. Living as they do in cramped apartment blocks (by North American standards), the public life of most professionals is played out in offices, restaurants and hotels. In this context, wealth and success tend to be expressed situationally in certain stereotyped scenes. Riding in a luxury car, drinking the right cognac in an upscale club, and using a cellular phone while in a restaurant are all signs that one is in the thick of things and moving in the fast lane.

Modernity

Hong Kong consumers are very well disposed towards new things. Novelty is virtually synonymous with improvement and progress. Research has consistently demonstrated that achieving a reasonable trial level for many new products is not a problem. What is difficult is getting repeated usage and brand loyalty.

Value-oriented

With increasing affluence, consumers are willing to pay more for products with more value-added features such as good design. Australia, for example, sells a lot of kitchenware to Hong Kong because it concentrates on imaginative designing and does not try to compete with PRC products on price.

Leisure

Hong Kong is a highly stressed society. No one ever seems to have enough time to do all that must be done, making people disinterested in any product which does not save time. Products that succeed sell leisure and relaxation.

THE RETAIL ENVIRONMENT

For exporters of consumer products, Hong Kong's main retail players are the chains, the big department stores and the thousands of small Mom & Pop stores. As monthly rents skyrocket beyond HK$500 per square foot, the most competitive retailers are those making the most cost-effective use of their shelf space. The others are either gobbled up or forced to strike up strategic alliances with better organized partners.

Background

Western-style retailing is a relatively new phenomenon in Hong Kong. Twenty years ago, Hong Kong's shopping facilities were traditionally Chinese and relatively unsophisticated. Supermarkets were few in number and most of Hong Kong's food supplies were sold in open-air markets lining certain streets.

At that time, retailing was very much a seller's market. Western consumer products were few in number and prices were high because of low availability.

Hong Kong's retail market, especially the retailing of food products and fast-moving consumer products (FMCP), has undergone profound changes in the past decade as a result of the rapidly changing lifestyle of Hong Kong consumers. The increasing number of giant housing estates in particular has greatly accelerated the development of Western-style merchandising. The number of two-income families is rising; women are finding it more difficult to get the shopping and cooking done. Processed foods are, therefore, in greater demand and people are ready to pay a premium for good lighting,

decent hygiene, sensibly-arranged stock and quick, courteous service. Food sales are responsible for 22% of total retail sales, which themselves amounted to 36% of consumer spending in 1990.

Supermarket and drugstore chains

Hong Kong's FMCP market is dominated by two supermarket chains (Wellcome and Park'N Shop), two drugstore chains (Mannings and Watson) and one convenience store chain (7-Eleven), all of which are controlled by one of two companies: Hutchison Whampoa Limited and Dairy Farm (an arm of the Jardine Matheson Group). Minor chains include the China Resource Group supermarkets which handle 23% of the Hong Kong-China trade, and the convenience store chain Circle K.

The extraordinary growth of these chains during the 1980s was accomplished at the expense of Hong Kong's thousands of poorly organized Mom & Pop stores. Their plans for the next five years include more and larger stores in the New Territories where the population is expected to grow, the creation of supermarket chains in Guangdong Province (40 stores are planned for 1994 by Park'N Shop alone), tighter inventory control through the computerization of checkout counters, and a push for greater market share in fresh and refrigerated foods.

From a distributor's perspective, the chains' stranglehold on Hong Kong's FMCP market raises certain entry barriers for new products. First, the chains have trimmed the number of distributors they deal with in an effort to reduce costs and rationalize their supply channels; all that remain are the best organized distributors such as JDH, Edward Keller and EAC (see Annex 3). Since the chains are experts in the most profitable use of limited shelf space and keep careful records of the turnover rate of each product, the distributor's value-added service to his suppliers is increasingly driven by information and technology rather than by the traditional pushy sales approach.

Secondly, the chains are in a position to impose onerous conditions on suppliers of new products. Once a product has gone through their rigorous selection process (new products can only get in by displacing less satisfactory items), suppliers are commonly asked to pay a one-shot listing (or slotting) fee of between HK$50,000 and HK$600,000 for access to shelf space. Other conditions include bigger "introductory" margins (5-10% more than usual) during the first 3 to 6 months, exclusivity, and a 60 to 90 day payment period instead of the 45 to 60 days asked by independent stores. They may also ask for compensation if their main competitor sells the supplier's product at a lower price. For these reasons, some suppliers systematically avoid the big chains.

Department stores

The main players in this market are the old-style Chinese chains, such as Wing On and Sincere, and some very aggressive Japanese retailers, such as Jusco and Yaohan which entered the Hong Kong market during the early 1980s.

Wing On is an excellent example of the difficulties many Chinese family businesses face when they are taken over by a second generation of management lacking the drive, the skill and the unchallenged authority demonstrated by the founder (see Management Hong Kong-Style). Since the death in 1986 of its founding manager, Dr. Albert Kwock, Wing On has lost most of its young customers because it has not been able to keep up with the changing times. To retain market share in the face of Japanese competition, Wing On has been forced to seek a strategic alliance with another Japanese retailer, Seiyu Ltd., part of the Seibu Saison Group and the owner of the Seibu department store chain.

The Japanese retail formula of bigger stores which incorporate restaurants and space for children to play has proven very popular in Hong Kong. With 80% of their merchandise locally sourced, some distributors feel that Japanese stores can be a good channel to introduce foreign products to Hong Kong.

Mom & Pop stores

The 1980s saw a dramatic decline in the number of Mom & Pop stores in Hong Kong, the principal reason being the success of convenience stores such as 7-Eleven. The Mom & Pop stores that remain are either well-located or better run than the average.

The majority of such shops cover 20-30 square meters and order the products they carry in frequent, small quantities. Most distributors, therefore, rely on wholesalers specialized in "capillary" distribution to reach them.

INDUSTRIAL AND GOVERNMENT PURCHASING

Industrial and government purchasing is usually conducted in a highly competitive bidding environment. When middlemen are used, suppliers usually rely on agents or manufacturers' representatives.

INDUSTRIAL BUYERS

Tough

Selling capital equipment to Hong Kong's light industry manufacturers is a relatively straightforward if not always profitable proposition. Hong Kong's free port status and the eagerness of foreign machinery sellers to establish market share in an area they perceive as the trendsetter for future purchasing in South China have pushed prices down to the point where profits are sometimes non-existent.

Criteria

The Hong Kong buyer has four selection criteria. These are, in descending order of importance:

- The supplier's **brand name** and **track record** in Hong Kong. For suppliers new to Hong Kong, particular attention has to be paid during the distributor selection process to the strategies various candidates propose to overcome this barrier. Some candidates may already be involved in manufacturing or be able to sell one's machinery as part of a larger package which includes items

more familiar to Hong Kong buyers.

- **Quality control reports** This is where the Europeans have a competitive advantage over most North American suppliers. Instead of waiting for an order before producing a machine, the Europeans produce their machines early and test them individually in order to have performance reports for each machine available to potential buyers.

- **Price** The pay-back period considered necessary in Hong Kong (4-5 years) is considerably longer than what is required in the Pearl River Delta (2-3 years).

- **After-sales service, warranty period, cost of spare parts, etc.** Unlike the PRC, preventative maintenance is common practice in Hong Kong.

Payment terms

Payment is usually staggered: 30% when the contract is signed, 60% when the machine is installed and 10% after a trial period (usually one month) which must be negotiated.

In some cases where competition is particularly stiff, financing arrangements may be required which allow the buyer from 180 days to 5 years to make payment.

GOVERNMENT PURCHASING

Who decides

Selling to the Hong Kong government requires that potential suppliers first be pre-qualified with the Hong Kong Government Supplies Department (GSD). GSD has the mandate to procure, store and distribute supplies required by the Hong Kong government and some related organizations. Accordingly, it has three divisions: administrative, procurement and general. The Administrative Division is responsible for general management; the Procurement Division handles purchasing and tendering; and the General Division deals with storage and internal distribution. It is the Procurement Division which is important for foreign exporters.

Invitations

Invitations to open tender are communicated through the Government Gazette, leading Hong Kong newspapers and foreign trade commissions in Hong Kong. While time limits may vary, six-week deadlines are quite common. Submissions should have a minimum validity period of 30 days.

Procedures

While the GSD usually purchases by open tender, selective tenders and single tenders are occasionally used when specialized commodities are being purchased or when security considerations mean that there is only one known registered supplier. In cases where purchases are valued at less than US$150,000, government departments need not use the GSD channel.

Selection criteria

Tenders are usually awarded to the lowest bidder meeting the requirements unless another supplier has a better overall package in cases where services or delivery are particularly important. In either case, unsuccessful bidders submitting offers lower in price are advised as to why their bids were rejected. Total bid prices and the names of successful tenders are published.

Who to contact

For more information, exporters should send inquiries to the following address:

The Director of Government
   Supplies
Government Supplies Department
The Hong Kong Government
12 Oil Street, North Point
Hong Kong
Tel: (852) 802-6102
Fax: (852) 807-2764

HONG KONG'S
DISTRIBUTION SCENE

DEFINING RELATIONSHIPS

North American manufacturers investigating Hong Kong-based distributors are often confused by the loose way certain trade terms are used. Although we will refer to "distributors" and "agents" throughout this book, it is important to note that these terms describe relationships rather than organizations. A given firm may, in fact, act as a manufacturer's representative for one supplier and yet be a distributor for another.

The trading companies

Most foreign exporters of commodity goods and products with low-level brand recognition depend on Hong Kong's thousands of local trading companies to penetrate the market. Trading companies typically match buyers with sellers on a commission basis, distribute the goods that arrive in Hong Kong, and assume neither financial nor credit risks. While some are large, wealthy and well-established, others may be new to the business or one-man operations. Some specialize in specific geographical markets; others trade only in a small number of products. Others may be controlled by or have ownerships interests in a particular wholesaler while some may be sufficiently independent to find the best wholesaler for a given product. A trading company serves a client in various capacities depending on the required transaction, its role changes depending on trade trends and its arrangements with individual suppliers. A supplier must therefore assure himself that the business relationship he seeks is made clear to any prospective trading company at the outset to avoid misunderstandings later. It is also worth noting that large wholesalers and supermarket chains, such as Wellcome and Park'N Shop, bypass trading companies by importing food items and certain fast-moving consumer products on their own.

Advantages

- Handles import documentation and provides short-term warehousing.

- Assumes the cost of physical distribution to the wholesaler.

- Saves the supplier the time and expense required to find a buyer for his product.

- May be a useful intermediary for suppliers wishing to use the "underground" option to test-market their consumer products in the PRC (see Market Entry and the Hong Kong Middleman).

Disadvantages

- The supplier loses control over pricing and thus over long-term market share.

- Brand identity remains undeveloped.

- The supplier has little or no communication with the Hong Kong customer, and cannot react effectively to market trends and opportunities.
- Because they are deal-oriented, trading companies are rarely interested in developing markets for new products or in providing after-sales service to end-users of capital equipment. Claims to the contrary in a given deal usually mean the trading company intends to subcontract this work to someone else.

The agent

An agent is a person or firm that is granted a carefully circumscribed right to sign sales agreements in the supplier's name in order to secure orders. Accordingly, agents rarely keep an inventory except occasionally on consignment, and bind their suppliers contractually as long as they do not overstep the limited authority originally granted to them. It is important to note that such agents are sometimes referred to as **stockists** if they are exclusive agents or as **dependent distributors** because they do not buy or sell on their own account. A great source of confusion is the incorrect application of the term "agent" to all trade intermediaries, i.e., distributors, sales representatives, manufacturer's representatives, trading companies, etc.

Advantages
- In matters of marketing strategy, pricing and sales methods, a supplier has more control over his agent than he has over a trading company or manufacturer's representative.
- Agents allow a supplier to get increased feedback from Hong Kong end-users, wholesalers and retailers.

Disadvantages
- Although this is open to negotiation, most agents do not assume credit risks; the supplier must investigate the credit-worthiness of every customer.
- According to Hong Kong law, a local agent may render his supplier liable for certain taxes depending on whether he negotiates and concludes sales contracts or simply solicits orders (see Taxation in Hong Kong).

ABC, a subdivision of a large multinational firm, among other things, specializes in the marketing of Italian shoes and leather products. To secure orders, ABC's sales team regularly visits department stores and boutiques selling luxury goods to investigate what is successfully selling and, at the same time, to convince the owners to stock up on their suppliers' shoes and handbags.

To ensure the sales force is motivated and well-informed, most of ABC's suppliers have agreed to pay the cost of sending product managers to Italy for a thorough introduction to their manufacturing and design processes. However, the suppliers insist on regular reports from ABC on its sales activities and on Hong Kong market trends. They also exercise tight control over ABC's marketing activities. Brand image is so important that much of the materials used in advertising come from Italy.

ABC does not buy the Italian products it sells. It consolidates and transmits a multitude of small orders to its suppliers and takes a commission on everything that is sold. ABC keeps no inventory except on a short-term consignment basis.

The manufacturer's representative

The manufacturer's representative is an independent person or firm that maintains a stable of suppliers specialized in a narrow range of products. Some are so small that they merely arrange sales and facilitate the flow of information between supplier and purchaser. Others, however, bid for government contracts or participate in one-shot deals connected with infrastructure projects, military hardware or heavy equipment. Manufacturer's representatives are like agents in many respects, except that they have no authority to sell goods in their suppliers' name. They are paid on commission and do not buy from a supplier on their own account or maintain an inventory. They may or may not provide after-sales service.

Advantages
- For small markets, such as Hong Kong or in fields where the product is bulky, expensive or made to order, manufacturer's representatives may be the logical way to successfully export .

Disadvantages
- Suppliers have very little control over the sales methods and marketing activities of their manufacturer's representatives.
- The supplier is still responsible for assessing the customer's credit worthiness.

Originally founded thirty years ago by a British entrepreneur living in Hong Kong, MZ now belongs to a Singaporian ship owner and is managed by two Americans. MZ has exclusive agreements with over 40 American and German suppliers of sports and park equipment and facilities. Its main clients are various branches of the Hong Kong government: Urban Council, Regional Services and the Hong Kong Housing Authority.

Like any construction firm, MZ bids for government projects by packaging its suppliers' products and its after-sale service warrantees into a competitive proposal. Ideally, its key sales staff might even influence the architect's original specifications to better fit its suppliers' products. Only when the contract has been secured will the suppliers be contacted. Storage costs are kept to a minimum by insuring that deliveries are timed to fit in with construction schedules. Payments are made to MZ which then pays its suppliers after taking its commission.

The distributor

The distributor is an independent company which takes ownership of its supplier's merchandise, maintains an inventory, and deploys its own sales, marketing and after-sales service staff in passing the goods down the distribution chain. In Hong Kong, they are sometimes referred to as **stockists, dealers, importers, designated agents, agents, brokers, representatives** or **sales representatives.**

Advantages

- Because a distributor buys and imports on his own account, suppliers need only concern themselves with his credit standing, not that of his down-line wholesalers and retailers.

- Using a distributor limits the supplier's exposure to tax, product and warranty liabilities in Hong Kong.

- Distributors are more capable and willing than agents or manufacturer's representatives to provide warehousing, servicing and market intelligence.

- Compared to trading companies and agents, distributors are far more willing to invest in market development work for new products and after-sales service for capital goods.

Disadvantages

- Because distributors are independent foreign businesses, suppliers often have less control over or even information about their distributor's down-line intermediaries, marketing methods and pricing practices than they would have with an agent. Supplier control depends on what has been agreed to in the distribution agreement and the degree to which the supplier's support helps the distributor earn money with that product line.

- The support required by most distributors to push a supplier's products in Hong Kong and especially in South China does not come cheap. In addition to the travel and legal expenses required to locate a distributor, do a credit check and make a contract, success often requires sales training, follow-up visits, promotional support and sales materials.

B Medical Equipment Ltd.'s core business centers around the 45 suppliers with whom it has exclusive distribution agreements. B also has non-exclusive agreements with over 100 other suppliers, generally manufacturers of low-value disposables and "me-too" products, but treats them as a sideline because their revenue-generating potential is limited.

B's customers are Hong Kong's 36 government-funded hospitals, 10 private hospitals and the many small clinics run by general practitioners.

All are very conservative buyers. B is therefore searching for suppliers with a proven track record and brand-name products, suppliers willing to invest in the detailed clinical evaluations and specialized seminars required to convince hospitals to switch brands. Long-term contracts with suppliers are also crucial because the procurement cycle for medical equipment is 12 to 18 months long. B's main contribution in such an agreement is the influence it can exert over its contacts during the bidding process, and its after-sales technical support which is crucial for hospitals.

The sales representative

Unlike an agent, who is retained by a group of suppliers, a sales representative is usually the employee of a single supplier and represents only his employer. As a result, most of his work revolves around selling: paying regular visits to existing clients, relaying orders to head office, finding and investigating the credit worthiness of new clients, informing his employer of market trends, etc. In time, and if his office is being upgraded to a sales branch, his job may orient itself around the supervision and support of third parties, such as agents and distributors.

Advantages
- Motivation. One or two of the supplier's own employees working full-time selling a product are

worth a dozen distributor salesmen who feel no personal connection with the supplier or his products.

Disadvantages
- Higher cost. Suppliers choose to go though distributors for a simple reason: the level of available business is not substantial enough to justify the high initial cost of setting up their own organizations.
- Certain markets, such as the PRC, require networks and a level of expertise very few sales reps possess. Many suppliers, therefore, find it useful to maintain long-term relations with their distributors.

MLM

Multi-level marketing (also called network marketing) is a form of direct marketing which relies not on commercial distributors but on private individuals who both sell the product and sponsor other people willing to do the same. In this way, by a process of duplication, a "downline" of distributors is created who sell by word-of-mouth promotion. The best known example in Hong Kong is Amway. Generally, MLM companies specialize in personal care and household products.

Advantages
- The concept of network marketing has proven very popular among Hong Kong Chinese because many aspire to run their

own independent businesses. One recent example of their receptiveness to MLM is Nu Skin International which signed up over 20,000 distributors in its first 10 months of operation in Hong Kong.

• Most of the marketing risks are assumed by the independent distributors, not the suppliers.

Indenting

Indenting is a British term describing an activity, not a relationship or an organization. To "do indent" or "sell on an indent basis" is to sell on commission without ever taking title to the goods. Accordingly, it might seem at first sight to be a transaction restricted to agents and manufacturer's representatives. However, the term is also used by Hong Kong distributors and even sales representatives if they receive a commission or bonus on a given sale where the merchandise goes directly from supplier to purchaser. Indenting is, therefore, used by a wide variety of intermediaries because it frees those securing the order from having to buy, import and resell goods.

TYPES OF DISTRIBUTORS

If, after careful study, a supplier is convinced that his products have a large enough market in Hong Kong or southern China, the next step is to select the best distributor or agent to do the job. Distributors in Hong Kong can be divided into groups, each with their own strengths and weaknesses.

THE CONTEXT

A little history

In the last century, when tea was China's chief export, the fortunes of British trading houses (*hongs*) such as Jardine, Matheson and Co., were tied to opium and tea leaves.

By the mid 1970s, these *hongs* had diversified into shipping, airline and aircraft maintenance, property development, public utilities and a host of other businesses such as hotels, dockyards, wharfs and supermarkets. The network between the four major *hongs* (Jardine and Matheson, Swire, Hutchison, and Wheelock Marden), the Hongkong and Shanghai Banking Corporation (renamed the Hongkong Bank) and senior levels of government was so tight that it was said that "power in Hong Kong resides in the Jockey Club, Jardine and Matheson, the Hongkong and Shanghai Bank and the Governor – in that order."

New pressures

Much has changed since then. The *hongs* are becoming more Asian in their management style. New competitive pressures are also forcing them to find new niches in a region that relies less on Western technology and investment than it used to.

• Hong Kong's burgeoning manufacturing and property development industries opened up two avenues in the 1970s where Chinese entrepreneurs could escape the domination of the colony's British commercial elite. The best known examples are the K.S. Li Group, which now controls Hutchison, and the Y.K. Pao Group. Even the *hongs* which have remained foreign-owned are now extensively locally managed.

• Hong Kong's impending reversion to China and the obvious leverage other Asian trading houses, such as Japan's *sogo shosha* or Malaysia's Sime Darby, gain by having a strong government-supported home base has pushed many *hongs* to seek closer ties with Beijing and Guangdong Province. In a world trade system increasingly dominated by trade blocks, Hong Kong's future is obviously tied to that of China.

• The rise of Western manufacturers with the clout to do their own marketing and distribution has long been a threat to the *hongs*. Now with Asian companies becoming more international, the pressures to offer a comprehensive, low-cost service are greater than ever.

- Products too have changed. Twenty years ago, when Hong Kong was a very different place and business was done at a much slower pace, many *hongs* accumulated the distribution rights to a wide variety of goods in an effort to become one-stop shops for everyone. With so many products to sell and so little time to become acquainted with them, their salesmen acted as order-takers rather than marketers. This approach no longer satisfies a growing number of suppliers whose technically sophisticated products have increasingly shorter life cycles. Such products can only be profitable if they are actively sold by a trained sales force focussing its efforts on clearly defined market segments. This new reality is forcing the *hongs* to specialize and become much more discriminating when adding new lines.

- Supermarket chains, such as Park'N Shop which do their own importing, combined with versatile smaller trading companies are pushing the large *hongs* out of the market for certain product categories where the intermediary's only role is to buy and sell. Because of their size and high overheads, large distributors are concentrating on product categories that require value-added services, such as financial support,

an information-gathering infrastructure, and after-sales maintenance.

Strategic responses

To maintain their competitive position, Hong Kong's large distributors have adopted certain common strategies:

- They strive to increase their profitability through specialization and larger, more controlled market shares. When done well, this also increases supplier and customer loyalty.

- Within the limits of this trend towards specialization, they seek to diversify their product range to avoid overdependency on a small number of suppliers or clients. A distributor already selling consumer products to the large supermarket chains, for example, will strive to find products of interest to the independent supermarkets.

- They develop a network of offices in Asia, particularly in the PRC, that manufacturers wishing to do their own distribution find too expensive to match.

- They offer potential clients a total package service including shipping, insurance, market intelligence, finance, distribution and customer service. This makes it less attractive for big exporters to rely on their own resources.

- They master skills unavailable to non-specialist traders, such as futures and countertrading.

- They participate in deals that tie in promising suppliers for a long time. One way to achieve this is to increase the number of joint ventures with suppliers to manufacture products that were formerly traded. This approach is very compatible with the PRC's own trade and investment priorities which emphasize the desirability of export-oriented and import-substitution projects.

TYPES OF DISTRIBUTORS

Transnational distributors

Examples of transnational distributors are the Inchcape Group, The Jardine Group, and The East Asiatic Company (see Annex 3). Most have their roots in Europe; however, a certain number have since been partially or wholly acquired by Asian interests.

From their early days as traders on the China coast, many transnationals have since become highly diversified conglomerates. In addition to marketing and distribution, many are active in shipping, insurance, retail trade, property development, and manufacturing. From a supplier's point of view, working with transnational distributors has certain advantages and disadvantages.

Advantages

- Well-established and financially strong.

- A large pool of foreign and Western-trained managers who know the market well.

- Easy communications because they understand and practise Western methods of doing business.

- Better able to distribute a given product to several Asian markets simultaneously because of their network of branch offices.

- Good bargaining clout with the supermarket chains.

- Some of them provide good after-sales service in the PRC.

- Easy access to sales information and the sales force for training and incentive programs.

- Better able to coordinate large-scale projects involving many suppliers.

Disadvantages

- Some suppliers find them bureaucratic and unaggressive. For niche markets, in particular, they are often too slow to react effectively.

- Only a small percentage of the thousands of products they carry are taken seriously because they are the "locomotives" that pull the rest along. Some suppliers also have a great deal of clout with them because their brand-name products are known worldwide. Suppliers not in that class are easily forgotten.

Large Hong Kong traders

These organizations can also be very diversified; however, as a rule their management style is far more centralized than that of the transnationals, as is often the case for Chinese family businesses (see Management Hong Kong-Style). For most Western suppliers, this makes them more difficult to understand.

Advantages

- As a rule, they are more specialized than the transnationals and carry fewer product lines. This can give smaller suppliers more clout.
- Many are very well connected in Hong Kong and the PRC.
- They are often more entrepreneurial and less bureaucratic than the transnationals.

Disadvantages

- Suppliers sometimes find them more secretive than the transnationals. Access to the sales force is less automatic.
- They are sometimes suspicious of Western-style marketing. Typically, they rely on their network of business contacts for information and promotion. Suppliers hoping for sophisticated market intelligence may be knocking at the wrong door.
- They rarely have networks of offices in the PRC that can match those of the transnationals. Things like after-sales service may be subcontracted to others.

- Because they are deal-oriented, quantity often tends to overshadow quality. So do short-term goals as opposed to the long-term. Few resources are therefore set aside to properly launch products or to build and maintain market share. Once a product has difficulty, they tend to drop it and move on to something else. Close supervision by the supplier is often required.

Major Hong Kong manufacturers

Although most of Hong Kong's manufacturers are export-oriented, a certain number have built extensive distribution networks within Hong Kong and South China. For suppliers whose products are compatible with those of such a manufacturer, i.e., if their products do not compete with each other but are intended for the same type of customers, the option of piggybacking on an established distribution network should be investigated.

Advantages

- They are often very aggressive and entrepreneurial compared to the transnationals.
- Within their own narrow speciality, these manufacturers may have as good a network of offices within the PRC as any transnational.

Disadvantages

- Like the large Hong Kong traders, most of these manufacturers are Chinese family enterprises. Western suppliers may therefore find them too secretive and disinterested in marketing.

- They are likely to give priority to their own products at the expense of the supplier's. Every possibility of product conflict should therefore be investigated by suppliers interested in this distribution option.

The rest

Smaller versions of those listed above may be worth considering for exporters new to Hong Kong whose products are relatively unknown. Once their products are well-established, they can then switch to a larger distributor with more resources.

Advantages

- Some small distributors are highly experienced and well established.
- They are very hungry for business and are likely to give the supplier's products close and individualized attention.

Disadvantages

- Finding good, small-sized distributors requires more thorough research from suppliers. Many small firms make inflated claims that are impossible to verify.
- Most of these distributors have too few resources to commit themselves to a product that does not take off quickly. Products requiring after-sales service should also not be entrusted to them.
- Small distributors frequently require more training, promotional and inventory support from their suppliers than larger firms do.
- Unless a supplier simply wants to test-market his products in the PRC by exercising the "underground option" (see Market Entry and the Hong Kong Middleman), he should not expect a small distributor to be able to effectively market his goods in China, regardless of any claims to the contrary. Small distributors do not have the financial resources and personnel required to do a good job.

MANAGEMENT HONG KONG-STYLE

Hong Kong's business world is actually several communities functioning in one territory according to different cultural norms. The engine room below deck, however, is the Chinese family enterprise, a business entity whose management style is at the heart of Hong Kong's business ethos.

PARALLEL WORLDS

The old stock

Hong Kong's first international entrepreneurs were not Chinese at all; they were the British and European merchant adventurers who made this "barren island" their base in the early days of trade with China. The trading firms they created, which survive to this day, have since diversified into very high profile conglomerates whose management style has played an important role in shaping Hong Kong's business culture.

Once family enterprises, most have long since gone public; their management is now known as professional and usually high in calibre. These same managers, however, have remained almost exclusively expatriate which has sometimes earned them a reputation for cultural aloofness by hiring through "old boy" networks, sticking with people of their own kind, and adopting a paternalistic attitude toward the local staff (about 95% of the total). This attitude of *noblesse oblige* has also made them easy marks for leftist union organizers whose real strength now lies with the old British firms rather than with young Chinese ones.

The new stock

The European, American and Japanese investors who flocked to Hong Kong in the 1960s and 1970s were much less concerned than the British with the way they were judged by the locals. Their main interest was in cheap labor and wealth extraction. Their factories were nevertheless relatively comfortable and efficient, and they generally applied scientific management principles with the local staff. American investment, in particular, opened up such key growth sectors for Hong Kong as plastics and electronics. While none of these foreign firms were interested in recruiting locals for top management, thousands of middle managers, some of whom now lead their own companies, learned their trade by working for them.

CHINESE FAMILY BUSINESSES

The context

Hong Kong's excellent communication and transportation infrastructure, its compact size and specialization in the production of light consumer goods after the 1950s (see Historical Overview) made it an ideal location for small and medium enterprises.

Light consumer products are fashion-oriented and subject to frequent changes in design and consumer taste. Flexibility in production and quick delivery to market are critical to success, giving a comparative advantage to small enterprises compared to large scale production by a few, big plants. But small size alone is not enough: small and medium enterprises in most developing countries are excluded from international trade by a host of negative attributes, such as low skill levels, low productivity, poor quality control, and lack of access to current information on production techniques, new designs and foreign markets.

Hong Kong's hundreds of thousands of small manufacturing and trading enterprises, on the other hand, are world famous for their rapid response to market changes, quick and punctual delivery, and constant improvements in quality and production techniques. What accounts for this dynamism? Since most small and medium enterprises are also family businesses, a good part of the answer rests in the way Chinese family values are applied to the workplace. The preeminence of men over women and age over youth, the abiding respect for education, the sensitivity towards personal and group prestige all contribute to this dynamism. Indeed, the Chinese family

enterprise's productive efficiency owes less to technology than to its particular combination of small size, internal loyalties and expediency with outsiders. What effect does this have on labor relations and on the ability of the family business to collaborate with other enterprises?

Father knows best

Internally, Chinese family enterprises are characterized by a very hands-on management style coupled with a familial work environment where obedience and loyalty are rewarded with protection and support (i.e., quasi-family membership) by the owner-manager. Three things in particular distinguish its approach to management: an autocratic leadership style, personalism, and the importance of ownership.

Chinese businesses are typically owned by a founding father and tend to be smaller than their Western counterparts. Over 92% of all manufacturing companies in Hong Kong employ fewer than 50 people. The owner-manager is usually heavily involved in the day-to-day supervision of his organization's lifeblood, its productive efficiency. As a consequence, he exercises almost unlimited authority, rations information to his subordinates, does not tolerate opposition even if it is loyal, and has great difficulty in delegating authority to non-family managers, however capable.

This hard edge is balanced by the owner's personal attention to the welfare concerns of his loyal employees. Although labor relations are generally good, the glue which holds labor and management together is not loyalty to the company; rather, it is the sense of obligation each employee feels towards the owner for his many past favors, e.g., employment, promotions, generous New Year bonuses, financial support for medical and educational expenses, soft loans, introductions and references, or even the simple act of showing interest in his personal affairs. These bonds remain important in small enterprises even if Hong Kong's tighter labor market has given workers increased bargaining power.

In a sense, these leadership qualities are the result of a fundamental belief that ownership is the only source of legitimate power and authority in a family enterprise and this has fateful consequences for management. On the positive side, reaction time is kept very short since the owner-manager can decide everything on his own and expect unquestioned obedience from his employees. This same authority, however, strips his middle management of any authority based on expertise or position. As a result, a decision is often based not on what one thinks is right but what one thinks the owner would do if he were present. This sets definite limits for talented middle managers who are not family members and must either break out or resign themselves to perpetual dependence.

Followers

The reverse side of the strong owner-manager is his obedient work force. Subordinates are less likely to take individual initiatives or volunteer opinions without a superior's approval. Opposition, when it exists, is voiced indirectly through third parties or expressed through passive resistance and a "go-slow" approach.

Another adverse side of the family management system is jealousy. Given the personal relationships between each employee and the owner, any evidence, real or imagined, that a given subordinate is favored by the boss immediately gives rise to the kind of jealousy one associates with siblings in a family. Resentments of this kind are very common since every enterprise has employees who are tacitly exempted from a normal work load because of seniority or nepotism. These employees are often feared because everyone suspects them of spying for the boss.

Favoritism greatly accelerates the creation of cliques or factions within the work place, something for which Chinese social life is notorious. Cliques can grow around a group of employees sharing a common birthplace, a common clan, or a common job (e.g., drivers). Because cliques are capable of placing their interests above that of the company, the owner-manager will do his best to nip them in the bud. Sometimes, however, their presence is institutionalized, as when owners allow gang leaders to bring in their own people and take care of a part of the manufacturing process for a fee.

Descendants

Given the Chinese family's distrust of outsiders and the fact that the owner-manager's closest associates are loyal to him rather than to one another or to the company, his retirement or death often precipitates the split-up of the organization into smaller units. This tendency is confirmed by Chinese inheritance rules which require family property to be divided equally among the children.

There is thus an in-built divisiveness in any Chinese family-owned firm. It is either fragmented when it goes through a generational changing of the guard or it grows to the point where the boss can no longer take care of everything. In the latter situation, he must hire a corp of professional managers with real authority to make their own decision. There is, however, a paucity of family enterprises that grow to this size or last beyond three generations.

The only exceptions to this rule, where Chinese-style centralized management can be made to work in large-scale companies, is when the firm confines itself to a single product amenable to centralized control, such as property development, chainstore retailing, shipping, hotel management or banking.

The entrepreneurs

A Chinese family enterprise is, first and foremost, an instrument for the accumulation of wealth by a very specific set of people. Short of marrying into the family, outsiders, however capable, can never achieve a level of authority family members take for granted. This creates a vicious cycle where talented managers break out to create their own enterprises, often after years of surreptitious moonlighting at the expense of their bosses. This, of course, confirms their employers' worst suspicions about outsiders and the stage is set for more break outs in the future.

There are three common scenarios in Hong Kong for these entrepreneurial break outs. First is the factory manager or engineer who knows the process thoroughly and decides to rent a factory, hire a few employees and buy some raw materials to try his hand at business. An even more common scenario is the marketing person who knows foreign buyers who care little about where they obtain their goods as long as the price is right. The salesman must simply find someone who will do the work more cheaply than his boss can, while leaving a margin larger than his salary used to be. The third scenario is the financier or promoter who entices a group of disaffected managers to leave their employers and help realize his project.

SURVIVING IN A NETWORKING SOCIETY

It is intriguing to note that Chinese family businesses, the most important players in Hong Kong's extraordinary economic development, have achieved their spectacular performance amid the context of chronic factionalism, rampant mobility and covert and overt disloyalty. How do such business entities, already distrustful of outsiders, go about networking with each other for mutual benefit? Part of the answer lies in Hong Kong's subcontracting system and the core values of its business class.

The subcontracting system

While Japan's industrial might is, in some ways, easy to understand with its large manufacturing companies and trading multinationals, the reasons for Hong Kong's export success are not equally apparent. After all, Hong Kong manufacturing companies have, on average, only 18 employees while trading companies have just six.

To some extent, confusion arises from our legal definition of the firm. We imagine a pyramidal structure with a CEO on top supported by functional departments (marketing, finance, production, etc.) in the middle, and, finally, an army of workers in factories at the bottom. The functional equivalent of a large Western firm in Hong Kong is a constellation of small firms whose components may change with each rush order and whose internal coordination is mostly based on prior ties and expediency.

From the individual firm's perspective, there are two kinds of subcontracting: developmental and commercial. While developmental subcontracting is common in Japan, it is rare in Hong Kong. Large firms, such as Toyota, seek long-term "parental" relationships with subcontractors where they not only get the components and industrial services they

want but also actively concern themselves with their technical, management and research development.

Hong Kong's commercial subcontracting system, on the other hand, is based on short term task-oriented relationships which provide more flexibility than the Japanese system at the expense of developmental activities. Because the export of light consumer products involves a multitude of small orders, seasonal variations and sudden rushes in demand, individual producers have difficulty planning production and are very cautious about expanding capacity. Dependability, however, is crucial and producers go to great lengths to establish a reputation for a quick and punctual delivery as well as a willingness "to make to order", regardless of the size and urgency of such orders. To do so, they must rely on the support of other factories in their network that have spare capacity. With a total of 280,000 firms in Hong Kong crowded into close proximity, almost unlimited combinations and permutations of capacity are possible.

Trading companies, which solicit orders from overseas buyers, play a vital role as contractors to a large number of small manufacturers that depend on them as marketing agents. Here, too, there is no correlation between a company's number of employees and its capacity to package short-term "consortia" to satisfy foreign buyers. In one study, a trading firm was found to have contracted the services of over sixty manufacturers and facilitated the creation of five enterprises, all in the space of one year and with a staff of 17.

Taken as a whole, therefore, this system is both flexible and powerful. Two other ingredients are also important: networking and the homogeneity of Hong Kong's business values.

Networking

Networking in a Chinese context is primarily the creation of a sense of mutual obligation through the exchange of favors. Networking, therefore, begins at birth when a child inherits not only a family with many extended branches but also its network of friends. Schools and university too create lifelong bonds between students in a class, bonds which have a special significance in Chinese culture.

Employment, of course, provides many more networking opportunities. Any business-related job gives access to resources – information, skills, hiring opportunities, etc. – with which one can accumulate IOUs from outsiders. The existence of cliques within the organization often enhances these networking opportunities. If a person helps someone find a job or gives him his first chance and subsequently trains him, that employee is now in his debt and will look after his interests. The existence of these internal loyalties are most visible when, for example, a marketing executive leaves the company to create his own firm, taking his team with him.

New companies and new ventures are often fertile ground for networking. Many rich Hong Kong investors like to take a small percentage share in all sorts of projects because of the access to new networks these investments provide. In the manufacturing field, a promising beginner will often find other manufacturers willing to lend him raw materials or pool spare parts in order to establish a solid basis for future cooperation.

Finally, Hong Kong is full of formal and informal groups which meet regularly to exchange information on problems of current interest. Many businessmen belong to clubs, associations and dining groups created around a single industry. These groups work best when they are small and include members whose sense of obligation to each other is sufficiently strong to loosen tongues. Often too they only include people from the same age group because communication across generations is not always easy in a hierarchical society.

Common values

According to Dr. Gordon Redding, author of *The Spirit of Chinese Capitalism*, horizontal linkages between Chinese family enterprises are facilitated by certain shared values, five of which are particularly important:

- **The work ethic** The narcotic effect of hard work, the respect its material reward inspires in others, and the great business value of a reputation for reliability impose on everyone a common way of seeing the world and evaluating others.

- **Pragmatism** The Hong Kong business person's almost exclusive focus on ends loosens inhibitions about means and facilitates the flow of interactions. Everything tends to be judged in terms of their practical outcomes.

- **The pursuit of money** The refugee mentality of many Hong Kong entrepreneurs (even though 60% of Hong Kong's population is locally born, 70% of its entrepreneurs were born abroad) creates strong pressures to achieve respect and security through the accumulation of wealth. In Hong Kong's seething money-making atmosphere, to stand still is to fall behind.

- **Insecurity** Hong Kong's older generation knew only war in China; the new one faces uncertain prospects after 1997. Everyone feels the pressure to maximize his revenue-making potential as early as possible. This malaise is exacerbated by the status anxiety of many up-and-comers, hence their fondness for name dropping, status symbols and prestige by association.

- **Fatalism and luck** There is a widespread belief that business is basically a gamble and that one has to learn to accept a great deal of uncertainty. Belief in fate and luck helps everyone rationalize the results, good or bad.

DREAM
OF THE RED CHAMBER

THE MAOIST LEGACY

China's pre-reform economic system and foreign trade regime are best understood as extreme examples of import substitution, a development strategy common among Third World countries during the 1950s. Its ultimate goal was to reduce China's dependence on foreign producer goods by manufacturing them internally.

BEFORE 1949

Economic regionalism

Although China's political culture has always emphasized the value of unity, its large expanse (9.6 million km^2), mountainous terrain, linguistic diversity and uneven population distribution (over 90% of Chinese live in the eastern and southern coastal areas, representing only 40% of China's territory) has meant that the country remained for millennia a mosaic of local and highly self-sufficient economies.

The countryside, where 80% of the population still live, was dotted with village communities whose economic and social activities centered around market towns. Typically, these towns served an area of 25 square miles (64.4 km^2) which included 8 to 10 satellite villages with a total population of a little over 7,000. Because the boundaries of these marketing centers did not overlap, each area constituted a self-contained community.

Four economies

Long years of warlordism, occupation by the Japanese and civil strife left China in 1949 with four relatively independent economies. First was the southern coastal sector centered around Shanghai where Western influence was strongest. This area mainly specialized in light industry, especially textiles. The second economy covered the vast interior and was based on self-sufficient traditional agriculture. Its intermittent contacts with Shanghai proved disastrous when the coastal textile mills destroyed its handicraft industry.

Heavy industry, built in the northeast (Manchuria) by the Japanese in the 1930s to supply their needs, was China's third economic pocket. That sector was not linked to the rest of China until 1949.

Finally, there was the pastoral economy of the ethnic minorities in Inner Mongolia, Tibet, Xinjiang and Qinghai. Like the northeast, this economic region was almost completely detached from the rest of China.

MAOISM: 1949-78

The CCP

Both its Leninist ideology and its revolutionary legacy predisposed the Chinese Communist Party (CCP) to see economic development as derivative of political objectives; there were no "economic decisions" as such, only political decisions with economic consequences.

Organizationally, the supremacy of socialist political values over economic considerations was ensured by the CCP's hundreds of thousands of party cells in every work unit of Chinese society. From the highest levels of government down to the factory and local committee levels, the Party set goals, controlled personnel assignments and served as a parallel communication channel, guiding local decisions in light of the priorities set by the party leadership.

Four players

From a distribution perspective, the centrally-planned economy which emerged is best understood by looking at its four key components as they relate to each other (see Fig. 4-1): the state production establishment, urban households, rural households, and the rest of the world (i.e., China's foreign trade system).

Figure 4-1

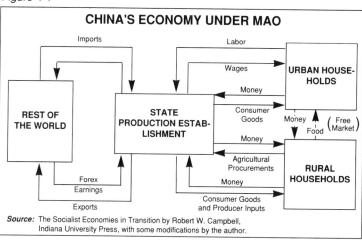

CHINA'S ECONOMY UNDER MAO

Source: The Socialist Economies in Transition by Robert W. Campbell, Indiana University Press, with some modifications by the author.

The first player, the Chinese **state production establishment** (government bureaucracy and state enterprises) was truly the core of the economy under Mao. Then, as now, it has a pyramidal structure with the State Planning Commission at its apex, ministries at the intermediate levels and primary production units (factories, hospitals, power stations, stores, ice cream street vendors, etc.) at the bottom. Though linked by strong vertical chains of command, this complex hierarchy also has large fault lines where intense bargaining rather than passive obedience is the rule. The most relevant fault line for distribution purposes is the tug-of-war between the national level on the one hand and the provinces, municipalities and counties on the other. Other fault lines exist between the "top-down" chains of command belonging to separate ministries. Authority and information in China's bureaucracy flows vertically rather than horizontally. If a disagreement over distribution arises between a national import-export (I/E) corporation and Shanghai's municipal retail department stores, for example, neither side has the power to dictate a solution to the other because they report to different ministries. If a compromise is not forthcoming, they must refer the question up their respective chains of command until it reaches a common boss.

The first priority for the CCP under Mao's leadership was the development of heavy industry (producer goods) and defence followed by light industry and agriculture. On the basis of these priorities, the State Planning Commission set one-year and five-year aggregate output targets which were passed down the hierarchy into more detailed production assignments for lower-level units. When an enterprise at the bottom of the chain received its output assignment, it determined the inputs it needed to reach that target and sent its requirements (i.e., how much coal, how many vehicles, etc.) back up the hierarchy until it reached the apex of the system. The State Planning Commission would then try to reach a "materials balance" by matching aggregate output targets with aggregate input requirements for the whole country. CCP priorities simplified this complex balancing process: the requirements of capital goods and defence factories were mostly accepted without dispute whereas light industry was treated as a "buffer sector" whose requirements could be trimmed to ensure an overall balance between aggregate inputs and outputs.

Urban and rural households, the second and third players in the economy, were linked differently to the state production establishment. Whereas urban households exchanged labor for consumer goods, rural households, collectivized into People's communes, exchanged agricultural outputs for consumer goods and such producer inputs as fuel and fertilizer. Given the overwhelming power of the state production establishment, both groups of households were "pricetakers" and had to make do with whatever state factories produced. On and off, rural households were allowed to sell produce from their private plots directly to urban households in "free markets". As a rule, however, such sales were curtailed and urban households relied on state stores for food at subsidized prices.

The rest of the world

China's relationship with the fourth player in its economy, the rest of the world, was regulated by a foreign trade system designed to insulate its centrally-planned system from outside forces. China's economic planners were inner-directed in their preoccupations.

In its **organization**, foreign trade became a state monopoly conducted through nine national import-export (I/E) corporations. The role of these corporations was to maximize the state's bargaining power in an international market where it was a pricetaker rather than a pricemaker.

The **role** of foreign trade was to help the State Planning Commission ensure a consistent materials balance, i.e., that output targets would be matched by input requirements. Developing heavy industry, for example, often required the import of advance machinery or raw materials unavailable domestically. That had to be planned for in advance because foreign currency was rationed. Exports, on the other hand, had no value on their own; they were perceived mainly, if not exclusively, as a means to finance imports.

Prices in China bore no relation to world-market prices and, therefore, did not serve as an incentive to trade internationally. Because producers were paid the same price, regardless of whether their goods were sold at home or abroad, and received none of the foreign exchange that resulted from foreign sales, they had little incentive to expand the production of goods for which there was strong international demand.

Many of the foreign traders remember the period from 1967 to 1972 – "before the Americans arrived and ruined everything" – as the most difficult of periods in many respects, but also, ironically, as the most profitable time to be selling to China. Outsiders were confined, spied on, unable to communicate effectively with the outside world, but their efforts were very, very profitable. "All you had to do was be able to put up with a lot of political nonsense till the last few days of the [Canton] fair. Then, Mr. Wang would throw you a piece of paper with the tonnage he needed to buy or sell according to the plan. That was your order, pretty much at whatever price you had quoted or bid weeks before," remembers a particularly successful European trader.

Source: *One Step Ahead in China: Guangdong under Reform*, by Ezra Vogel, p. 342, Harvard University Press.

China's **official exchange rates** highly overvalued the Renminbi (Rmb) to facilitate imports; however, the resulting losses on export products had no effect on the financial status of producers of export products or I/E corporations because profits from imported goods were used to offset export losses. Trade planning thus came to rely on a complicated system of cross subsidies and a stringent regime of foreign exchange control. The Rmb was inconvertible and all foreign exchange holdings had to be deposited with the Bank of China, the sole bank authorized to deal with foreign exchange.

Maoist politics

To this Soviet-style central planning system, Mao added certain features peculiar to China:

- After 1958, the provinces were asked to develop **self-sufficient economies** on the grounds of national security and because the central government could thereby concentrate more resources on the development of heavy industry.

- Provincial self-sufficiency meant that an effort was made to develop **industries in the interior** to redress the over-concentration of modern industry on the coast.

- Even with regional development policies for the interior, the crucial **income differentials** in China were and are still not interregional but between city and countryside. Because of government subsidies to urban households, the per capita incomes of town-dwellers since 1949 has, on average, been three times higher than that of rural residents.

- Material incentives for workers were de-emphasized in favor of such **spiritual values** as egalitarianism and class struggle.

- The CCP leadership, Mao in particular, was deeply influenced by the **military campaign work style** which had been their way of life for 25 years before 1949. Not making allowances for the more complex (more "selfish") needs of people in peace time, they felt that economic development was a war waged against internal class enemies in a hostile international environment dominated by imperialism, and that success in war depends on a clear chain of command, a single ideology, and bursts of common effort strategically directed. There was, therefore, a tendency for them to rely on political movements to weld "the people" and the Party into one invincible whole to speed up economic development. In reality, most of these movements – collectivization (1951-80), the Great Leap Forward (1958-60), and the Great Proletarian Cultural Revolution (1966-76) – proved disastrous.

Economic consequences

Mao died in 1976 and Deng Xiaoping assumed the leadership of the CCP in 1978. The consensus within the party at that time was that Mao's political movements and China's unbalanced economic development had seriously eroded the CCP's prestige among the masses. More specifically, agriculture had to be liberalized to boost production; about 10% of China's 950 million people did not have enough to eat. Secondly, more resources had to be devoted to the development of light industry to solve the chronic shortages of consumer goods in China's cities. If that meant unequal development for a while in favor of the coastal regions or a greater role for market forces, so be it, as long as the party's authority remained unchallenged.

REFORMS UNDER DENG: 1978-93

Deng Xiao-ping's reformist policies have enabled China to enjoy one of the world's fastest rates of economic growth over the past 15 years. In contrast to the former Soviet Union, where privatization of state enterprises and political liberalization were pushed to the fore, the most striking element of the Dengist reforms is their extraordinary dependence on foreign trade and investment as engines of development. A comprehensive description of China's tortuous reform process in a book on distribution would bring us too far afield. We will limit ourselves to the aspects of most interest to foreign exporters.

NEW PLAYERS

Despite repeated calls for accelerated change, the political, economic and social costs of reform were such that what has emerged is neither capitalism nor communism, but a mixed "semi-reformed economy" which is likely to remain a reality for many years to come. Compared to the old system, this new economy has a new cluster of players (see Fig. 4-2), the relative power of which varies greatly from province to province.

The **state production establishment** remains very important in heavy industry and regions where heavy industry is concentrated, such as the northeastern coastal provinces and Shanghai. There are over 13,000 large and medium state-run industrial enterprises in China, accounting for 45.6% of aggregate industrial output and providing over 60% of all taxes and profits paid to the government by industry. The main thrust of reform at this level is administrative decentralization in which decisions are moved down the administrative hierarchy closer to the enterprise. Decentralization, in turn, depends on Beijing's growing ability to replace direct commands or allocations in physical terms characteristic of planned economies with "economic levers" which attempt to steer enterprises through tax rates, subsidies, interest rates and bank credit. Decentralization of decision-making in state enterprises is nevertheless an uphill battle. Managers are enmeshed in a host of political and social obligations that constrain their room to manoeuvre, and most enterprises are overstaffed, use backward technology and turn out low-quality products that remain unsold in state stores. At least two thirds of these enterprises would go bankrupt were it not for "loans" (subsidies) from state banks which are never repaid. A majority of these state firms in the red belong to the defence industry and must now learn to produce competitively for the civilian market.

Figure 4-2

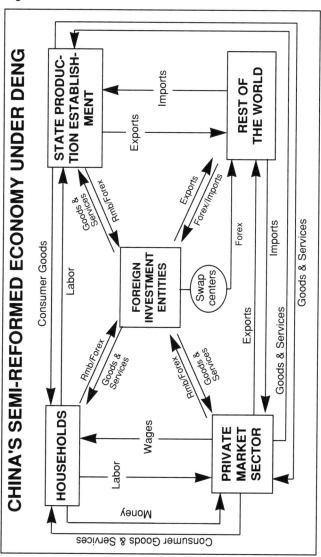

CHINA'S SEMI-REFORMED ECONOMY UNDER DENG

As long as private-plot agricultural production for sale on the open market is included in the private market sector, urban and rural **households** can be considered a single entity. Both are on common grounds vis-a-vis the state production establishment and private market sector.

The **private market sector** is a key new player, accounting for 69% of China's total employment. Its most visible members are farmers selling produce in the "free market" and small urban entrepreneurs (called *geti hu*), active in certain service activities such as repair, restaurants, and small scale retailing. However, this fast-growing sector also includes millions of rural "collective enterprises" established by local governments rather than by private individuals and, in some cases, the distinction between state and collective enterprises can be fuzzy. One final point: "private" in China often means "freer enterprise" not capitalism *per se*. The private market sector remains dependent on the state production establishment because the latter, in its role as government, controls access to scarce resources, levies taxes and issues permits. Indeed, the greatest profits are often to be had through the connections private enterprises carefully nurture with officials inside the state system.

Foreign investment entities include wholly foreign-owned ventures, equity and contractual joint ventures, Hong Kong-style cooperative ventures, and nebulous partnerships arising from compensation trade agreements and processing and assembly arrangements. As a group, these entities have been granted increasing access to the Chinese market and are becoming very important players in southern China's coastal provinces, especially in Guangdong. If we accept as a guideline the CCP's traditional attitude that what is not explicitly permitted is forbidden, many activities, especially the most profitable activities, of these entities fall into grey areas. China is full of grey areas.

BACK TO REGIONALISM

Decentralization has meant a resurgence of economic regionalism, a fact of Chinese life which never disappeared even under Mao's centralized command economy.

Agriculture

Since 1979, people's communes have been disbanded and replaced by a "production responsibility system" which allows peasant households to work their own land and sell any produce exceeding a certain minimum owed to the state on the free market. The result has been a dramatic increase in agricultural productivity and a partial return, especially in landlocked provinces, to the market town economies that prevailed before the revolution. The loosening of the state's control over rural society has also, however, revived old problems such as high birthrates, a deterioration in the systems for health care, education and water management, and the reappearance of folk religion, clan activity, gambling and gross discrimination against women. The rural responsibility system has also exposed the fact that as much as 50% of China's agricultural labor force, or 200 million people (a figure equivalent to one and a half times Japan's entire population) may be redundant. Due to reduced restrictions on the movement of peasants since 1984, over 100 million peasants have since moved into cities and townships looking for work. The "floating population" in some of China's largest cities is estimated to be 1.4 million in Shanghai, 1.2 million in Guangzhou, 1.2 million in Beijing and 1.1 million in Tianjin. With them has come a surge in black marketeering, smuggling, drug abuse, trafficking in imported commodities and prostitution. The kidnapping and sale of women and children, for example, has become a 100 million yuan business.

Coastal winners

The main thrust of the CCP's economic reforms in the 1980s was to connect the economy of China's south eastern coastal provinces with the market economies of the West through improved conditions for investment and freer trading practices. Between 1984 and 1988, industrial output in the four coastal provinces of Guangdong, Fujian, Zhejiang and Jiangsu increased more than 140% in real terms, four times faster than the slowest growing provinces. Preferential investment policies

and proximity to Hong Kong and Taiwan were key success factors for Guangdong and Fujian (see The Birth of Greater China). Jiangsu and Zhejiang have benefitted more from their proximity to Shanghai and the expansion of small-scale industry in their towns and villages.

Landlocked losers

Measured in terms of regional per capita income, China's interior provinces, such as Jiangxi, Henan, Sichuan and Hunan, experienced a short-term gain after the rural responsibility system was introduced and agricultural procurement prices increased between 1979 and 1983, but there has been no improvement relative to the national average since then. One reason for this has been the skewed structure of the price system which guarantees disproportionately more profits to coastal processing centers, like Shanghai, at the expense of producers of agricultural goods in the interior. Attempts to redress this situation led to "commodity wars" – the wool war, the leaf tobacco war, the silkworm cocoon war, etc. – during the 1980s where leaders in the regions erected provincial trade barriers to encourage more local processing and raise prices for outside buyers. Under the recessionary conditions of 1989-90, these protectionist measures were extended beyond raw materials to the local markets. The "China market" became a patchwork of rigidly separate local economies.

CHINA (PRC)

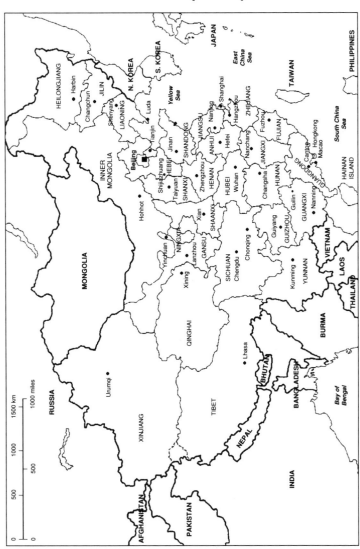

The starved center

Under a fixed "contract responsibility system" begun in 1988, a number of fast-growing provinces were permitted to retain all tax surpluses after remitting a fixed sum agreed upon with the central government. The center, however, is facing a funding crisis because the state enterprises which were its main cash cows when the economy was completely planned cannot function as well in a deregulated environment and must be propped up through "loans" from the state banking system which are unlikely ever to be repaid. The resulting deficits, by some estimates, total as much as 10-14% of total revenue. Beijing has no choice but to scale down central planning even further since it can no longer afford the heavy subsidies involved. More control over economic decision-making must be transferred to the provinces and to individual enterprises.

REFORMING THE FOREIGN TRADE SYSTEM

Decentralization

Decentralization has meant that the central government has relinquished its monopoly on foreign trade. In 1978, nine state I/E corporations handled all foreign trade. By 1990, the number of foreign trade corporations soared to over 5,000 because ministries, provincial governments, cities and enterprises all got in on the act. Ironically, this massive increase in possible trade partners made trade between foreigners and China more complicated, because I/E corporations did not all have equal access to foreign currency. This created a huge demand for intermediation in Hong Kong due to its familiarity with Chinese conditions and Western business practices. More decentralization in China meant more trade through Hong Kong.

Agency system

There has also been a shift towards an agency system for I/E corporations. Prior to reforms, national I/E corporations acted as principals buying from Chinese producers for export or from foreign producers for import; regardless, the responsibility for all profits and losses rested with the I/E corporations. Under the current agency system, I/E corporations only act as the agents for the exporting PRC producer or importing PRC buyer and get paid on commission. The implications of this shift for foreign exporters are important. Although I/E corporations execute distributor-like functions, they represent the Chinese buyer, not the foreign supplier. They will not therefore provide sales forecasts to foreign exporters, promote their products or provide after-sales service. They merely facilitate individual transactions.

New forms of trade

To promote exports, the central government introduced new forms of trade, the most important of which were export processing, compensation trade and cooperative ventures.

- **Export processing** includes both the processing of imported raw materials for export and the assembly of imported components to produce final goods for export.
- **Compensation trade** involves the supply of equipment or technology to a Chinese enterprise by a foreign firm in exchange for goods produced with that equipment (direct compensation) or other goods (indirect compensation or countertrade).
- **Cooperative ventures** combine elements of compensation, trade and joint ventures in projects where returns are more long-term. Usually the Chinese supplies factory premises, labor and raw materials while the foreign partner contributes equipment and capital. The Chinese manage the project and repay the foreign partner with products or cash.

The development of these three forms of trade was closely linked to the establishment of "open areas" which were three-tiered in terms of increasing degrees of autonomy: coastal open areas, coastal open cities, and special economic zones (SEZs). The key SEZs are Shenzhen, Zhuhai and Shantou in Guangdong Province and Xiamen in Fujian Province.

Fewer subsidies

In January 1991, the central government put a stop to the traditional system of compensating exporters for losses caused by an overvalued Rmb with the profits made by importers (see The Maoist Legacy). However, even with a gradual devaluation of the Rmb, many Chinese exports remain uncompetitive on world markets, and certain coastal provinces have decided to bail out their troubled export trade corporations by allowing them to import foreign consumer products for sale in the profitable domestic market to offset the loss of subsidies from Beijing.

Retained foreign exchange

Beginning in 1979, the central government gave up its monopoly on the control of foreign exchange, allowing exporting enterprises and local governments to retain a share of any foreign exchange earned from exports. Retention rates were last standardized in January 1991, leaving anywhere from 40-70% of foreign exchange earned from exports at the local level. This placed huge quantities of foreign exchange beyond the direct control of Beijing and was a major step towards internal convertibility of the Rmb for trade transactions. By 1988, over 40% of China's total export earnings (US$18.5 billion) was in local hands. Retained foreign exchange has also fueled more conflicts and trade barriers between the interior provinces, which produce raw materials, and the coastal provinces, which process and export. This is primarily because retained foreign exchange accrues entirely to the exporting firm and to the local government supervising its affairs. To redress this problem, in

1991 Beijing reduced the retention rate of such coastal SEZs as Shenzhen down to 50% and boosted the ratio for inland regions. Retention ratios will also be changed to favor Shanghai which, until now, has had to remit 75% of its revenues to the center whereas Guangdong only remits 10%.

Swap centers

The value of this locally-retained foreign exchange was enhanced considerably by the creation in 1985 of "foreign exchange adjustment centers" (swap centers) where authorized holders could sell foreign currency for Rmb at a more favorable rate than the official exchange rate. Since then, the numbers of swap centers has grown to almost 100 and total activity on the market has soared with China's surging exports. Of particular significance was the 1988 merging of

what had formerly been distinct swap centers for foreign and Chinese enterprises. This both increased the amount of currency available on the swap market and helped narrow the gap between the swap market and official exchange rates to less than 10% (see Fig. 4-3). A third rate, the black market rate, has also moved within 10% of the swap rate because of frequent devaluations of the Rmb, looser foreign exchange rules and the ineffective enforcement of China's custom regulations. (Some HK $40 million worth of contraband is smuggled into China from Hong Kong every night.) China is also the biggest foreign holder of Hong Kong dollars, with over HK$30 billion in circulation mostly in southern China. It is no surprise that the Hong Kong dollar and the Rmb are both in use at every level of Guangdong society. The Rmb

Figure 4-3

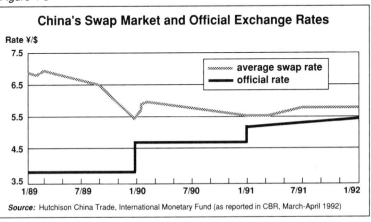

China's Swap Market and Official Exchange Rates

Rate ¥/$

- average swap rate
- official rate

7.5 | 6.5 | 5.5 | 4.5 | 3.5

1/89 | 7/89 | 1/90 | 7/90 | 1/91 | 7/91 | 1/92

Source: Hutchison China Trade, International Monetary Fund (as reported in CBR, March-April 1992)

85

is fast on its way to becoming a convertible currency.

POLITICAL EPILOGUE

Although the Tiananmen Square incident in June 1989 is frequently portrayed in the Western media as a political upheaval pitting idealistic students against an army controlled by conservative octogenarians, it is clear that economic factors were very important in providing a background to the uprising. Four factors in particular should be watched carefully by anyone thinking of making a long-term commitment to the China market.

Inflation

The large number of politically-aware groups in urban centers living on fixed incomes makes the governments of semi-reformed economies, such as Russia or China, extraordinarily sensitive towards the politically destabilizing influence of inflation. According to some estimates, as much as one third of urban households in Beijing were suffering loss of real income in 1988-89, and these included many of the civil servants, teachers, white collar workers and students who participated in the Tiananmen Square demonstrations. The CCP is aware of this but will not hesitate to reintroduce austerity measures similar to those it used in 1988-90 if the economy appears to be overheating again.

Income inequality

One of the legacies of thirty years of Maoism was an unwritten social contract between the regime and society which shaped the Chinese citizen's sense of justice. Socialism, for most urban dwellers, meant security (law and order), stability (lodging, medical treatment and staple foods at constant prices) and, in principle, if not always in practice, equality. All three of these pillars were removed during the 1980s, leaving many who once had been in comfortable positions losing out to taxi drivers, street vendors and smooth operators. They are not upset simply by the large amount of money these people make; it is the source of this income that enrages them. Semi-reformed economies inherit big armies, cartelized markets, weak legal infrastructures, multitrack pricing systems, overstaffed bureaucracies and inefficient tax systems. As a consequence, a greater amount of money can be made by preying on the irrationalities of the regime than by trying to develop better products for the market. The result is seen as "corruption" by the millions of ordinary party members and citizens unable to stay ahead in this new game, and corruption greatly undermines the CCP's claim to legitimacy. There is not much the Party can do about this except preach; already its labor camps are full. The two groups about which the CCP is particularly wary are university students and urban workers. Any serious disruption of public order from these groups could push the CCP to change the course of its economic policies.

MFN

America and China granted each other Most Favored Nation (MFN) trading status in 1980. Under US law, MFN status for countries with non-market economies must be reconfirmed every year by a presidential waiver, something which, in China's case, has become a very contentious issue in Congress ever since the Chinese government suppressed the democracy movement in 1989. An American decision to revoke China's MFN status would trigger a chain of retaliations that would devastate the US-China trade and severely affect Hong Kong's economy. Barring any dramatic improvement in China's human rights record or weapons sales policies, the debate over China's MFN status is likely to remain a key political issue in the US for the foreseeable future.

Deng's health

Of the twenty-five to thirty-five people who now make up China's top leadership, a core of four to seven highly-respected semi-retired leaders, including Deng Xiao-ping, set policy guidelines. With so much power in failing hands, the life expectancy of these elders remains a topic of intense speculation. Deng, now 88 years old, has virtually given up appearing in public while the other gerontocrats are said to be in poor health, even as they vie with each other to place their proteges in high positions. The order in which they die and the timing of Deng's exit, in particular, will greatly affect the jockeying for power within the CCP and, hence, China's economic policies. Investors and exporters interested in the Chinese market should therefore pay close attention to what develops in the year following Deng's departure from the scene.

THE PRC CUSTOMER: CAPITAL GOODS

As in most developing countries, business life in China revolves around a simple and inescapable fact of life: scarcity. Success in the PRC demands perseverance and an ability to capitalize on constraints.

A WORLD OF SCARCITIES

Energy

Many of China's fastest developing regions contend with serious power shortages. In the Pearl River Delta, for example, industrial growth and the popularity of electrical appliances, such as air conditioners among newly-affluent consumers, have meant that demand outstrips supply by over 40 percent. During the summer months, power shortages idle about 50% of Guangdong's industrial capacity. With government power supplies available only a few days a week, manufacturers risk losing orders if they do not install their own diesel generators. Even if Guangdong reaches its goal of having a per capita capacity of .2 kw by 1995, it will still be far behind Hong Kong and Taiwan which already have 1 kw per person.

Skilled labor

Except for highly industrialized urban centers such as Shanghai, skilled labor is scarce while unskilled labor is abundant. Literacy rates in the countryside, the main source of workers for SEZs such as Shenzhen, are commonly under 50%, lower still for women. This has obvious repercussions on the nature and value of any training offered to a buyer of capital goods: an extraordinary amount of time is spent in PRC work units with face-to-face meetings in order to communicate information and enlist support. Moreover, workers, once trained, often have more attractive opportunities elsewhere. With turnover rates between 20% and 30% in Guangdong, training must be seen as an ongoing function by suppliers looking for a good distributor.

Foreign exchange

This is an important factor at all levels, especially with state enterprises which must petition their ministries for foreign exchange and then, once funding is approved, collaborate with the relevant I/E corporation to select machinery. Indeed, one of the services a good Hong Kong distributor can provide is help for potential Chinese buyers in locating PRC enterprises or joint ventures willing to swap their foreign currency for Rmb. China's year-to-year budgeting and allocation system for foreign purchases encourages buyers to consider only up-front costs, not the cost of spare parts and after-sales service.

As a result, firms offering the best overall price often lose out to competitors with lower up-front prices.

Infrastructure

Weaknesses in transport infrastructure and services, telecommunications, and storage have several consequences. First, they lengthen distribution channels. Second, the poor quality of roads often causes more wear and tear on vehicles and a greater need for spare parts. Third, deficiencies in transport and storage frequently lead to sizable losses. A distributor may be asked to repair machinery which has been left to rust for months in a port or in the purchaser's back lot before being installed. Fourth, unlike Hong Kong's highly specialized small and medium enterprises which take advantage of the colony's good infrastructure by subcontracting jobs they cannot do to others (see Management Hong Kong-Style), China's state enterprises "solve" the problem of poor communications by becoming as vertically integrated as possible, incorporating every step of the production process under one roof. As a consequence, a machine's productivity may be far less important to them than its up-front price because machines remain idle 50% of the time.

Information

As in most developing countries, China's infrastructure deficiencies impede data collection. Mail and telephone surveys are much more difficult to implement. Moreover, cultural norms inhibit giving information to strangers, and the lack of trained interviewers makes much of the data easily available in the public domain unreliable. Finally, China's political tradition also limits access to information because access traditionally depends on one's position in the hierarchy. The CCP's style is to dole out information downward through official channels where it becomes increasingly general and less specific, more about ideological principles and less about practical considerations as it flows down the system. This style is mimicked in almost all PRC organizations, making information gathering a very personal, labor-intensive process.

COPING

On autopilot

Many PRC middle managers, especially those working in the more bureaucratic state enterprises, accept their powerlessness and learn to navigate within the system. They rarely take initiatives, responding only to written instructions where someone else assumes responsibility.

Little kingdoms

With so few willing to take initiatives, those who do reach top positions employ a largely autocratic style of decision-making with a minimum of formalized business procedures. Internally, the boss is king of a very fractious mountain; externally, much of his energy and time are devoted to negotiating bureaucratic deals by exchanging favors with his network of contacts in useful agencies. These networks of obligations sometimes become so demanding that a top manager may find himself powerless on many issues. The Western concept of the enterprise as an individualistic atom fighting for survival in an impersonal market environment corresponds to nothing Chinese managers have ever seen. State enterprises, like arms and legs, are part of something larger and must function accordingly. Private enterprises are indeed more independent and do fight for survival; however, the environment they function in is far from neutral. Resources are scarce and access depends on the exchange of favors.

The Chinese used to say, "One depends on his parents at home and his friends when travelling." Now, this friendship or network of contacts is not just necessary for travellers but a must for anyone wishing to run a private business in China.

"I run my own manufacturing business thanks to my friends. This is not a secret," said Mr. Wang in his crowded apartment, his grandson seated on his shoulders and his old mother beside him.

Three years ago, Wang decided, at the age of fifty-five, to retire from his "iron chair" (a guaranteed position for managers) in a state-owned factory in a southern city. "With my modest salary, I simply could not support my big family. We couldn't cope with inflation and we would

stay poor if I didn't do something on my own."

Thanks to his many long-time friendships, he managed to acquire a private loan of Rmb 40,000-50,000 to purchase some machines to start his own business.

When I asked how he obtained a valid permit and how much tax he pays, he laughed. "Frankly, I don't have to deal with all these things. I have the protection of a big "umbrella". Two of my associates are in charge of the service bureau for retired employees of my former factory. My company is registered under this service bureau. As long as we do everything in cash, I am OK, you know!" said Wang, winking at me. This is quite a contrast to the billboards extolling the virtues of blue uniformed workers one saw in the past.

"Now, I am much better off", continued Wang, now holding his grandson on his knees. "But, I never tell anybody how much money I make, not even my wife!" His happy face shone under the 60 watt bulb hanging in his living-room.

Millions of Rmb are kept in cash in people's homes because the Chinese fear the state-owned People's Bank of China, "the second bureau of security", as it is called.

Others enjoy similar favors as Wang. A middle-aged man I met while travelling to Shanghai revealed his secret: his private business, with an annual net profit of more than Rmb 200,000, is officially listed as the workshop of a high school. "This spares me all the headache of private entrepreneurs", he said. But I was even more surprised when this nouveau riche entrepreneur told me that he didn't have to pay for a train ticket. "The security man in this train is my friend," he added, showing off his large gold ring.

Consequently, those who lack connections have to face greater competition. A twenty-year old taxi driver, Xiao Liu, dreamed of opening a restaurant for years. "But, what can I do; I don't have connections!"

While we were caught in Shanghai's daily traffic jam, Xiao Liu explained that to quit his present job and start a private business, he must obtain permission from his state-owned taxi company and his neighborhood committee. Furthermore, he will need to obtain approval for his application from many government offices: the security bureau, the tax bureau and trade office, etc.

"Each required stamp means a lot of money or gifts; I just cannot afford this", said the young man, suddenly speeding up in downtown People's Square. "This is the only place where we can go as fast as we want." he said.

Source: Ms. Zhang Xiao-ling, Announcer-Producer, Radio Canada International.

Stepping over the line

Since PRC managers must create and nurture an extensive network of relationships to make the system work, it is only a small step for them to apply these same networks to supplement their meager salaries. Whether this amounts to corruption in the context of international trade depends on one's point of view; many of these managers wish to be paid for networking services which, in a North American context, would be paid to consultants, lawyers and accountants who fulfill the same functions. Corruption is more obviously carried out in requests for payments to improve business, bribes to push goods through customs or the Commodity Inspection Bureau, foreign trips with large per diems, sponsorship of education for trading officials' children, money to engage in black-market trade of import and export licenses, etc. This type of corruption is on the rise in China, more so in the south than in the north.

CUSTOMER PROFILE

Purchasers of capital equipment in southern China can be divided into four groups depending on their source of funding:

Government-funded enterprises

This group includes most of the state enterprises. A given factory fortunate enough to be in a priority sector or region as defined by the state plan will apply to its up-line ministry for foreign currency. If this request is approved, the ministry will inform an I/E corporation under its authority that amount X is set aside for factory Y to buy a certain type of machinery. It is only when the factory knows how much has been granted that it will begin to seriously evaluate purchasing options.

At that stage, the input of most I/E corporations is minimal because they lack personnel with technical expertise. The one exception to this rule is Techimport which handles tenders for large state-funded projects, turnkey projects or projects financed by the World Bank, the Asian Development Bank or foreign governments.

To access this market, distributors must follow state plans closely both at the central and provincial levels; the best ones are continually in touch with ministries, I/E corporations and factories to influence perceptions and specifications before the tendering and evaluation process even begins. Long-term familiarity with the buyers is crucial with this group.

Self-funding enterprises

Not all state firms are bloated dinosaurs. This group includes state and private firms which export successfully and have been allowed to retain a certain percentage – 40% to 70% in most cases, 100% for especially favored sectors, such as defence – of the foreign exchange derived from their sales abroad.

Everything said about government-funded enterprises with regards to being informed and maintaining contacts is equally true of this group.

Self-funding enterprises, however, are easier to identify than state-funded factories especially since January 1991 when the central government standardized the percentage of foreign currency earnings these exporters are allowed to retain (see Reforms under Deng). The purchasing process also tends to be simpler because decision-makers are fewer and control their own funds.

Equity joint ventures

In equity joint ventures, both partners have a say in the final choice. In most cases, the foreign party's technological preferences are respected by the Chinese side as long as the revenue projections originally agreed upon are not negatively affected.

As equity joint ventures are easy to locate and most of their foreign partners have representation in Hong Kong, the information-gathering and selling process is relatively straightforward.

Hong Kong ventures

This includes cooperative ventures, contractual joint ventures, compensation trade agreements, and processing and assembly arrangements. Most of the purchasing for this group is done in Hong Kong with the same selection criteria used for machinery purchased for domestic use (see Industrial and Government Purchasing). The one difference is that the payback period for capital goods purchased by this group is often even shorter than is required in Hong Kong because most Hong Kong ventures in

Guangdong are fairly short-term (2-3 years). Indeed, machinery is often bought on a project-to-project basis and must pay for itself before the project ends.

MOTIVATION

Since the selection of capital goods in joint ventures and Hong Kong ventures is basically determined by non-PRC "outsiders", we will concentrate in this section on the purchasing behavior of government-funded and self-funded enterprises.

Brand-ranking

In general, ministries, I/E corporations and factories are very conservative purchasers, preferring to stick with familiar suppliers and brand-name products with a track record in China. Those responsible for purchasing tend to have strong opinions, sometimes influenced by advertising, about which brands are best for every product category. Selling them new products and unfamiliar brands is not impossible; but it demands a lot of planning, a well-connected distributor and plenty of promotion through trade fairs and technical seminars (see Market Entry and the Hong Kong Middleman).

The party with the most say at the initial stages of the selection process, after foreign currency has been set aside, is the factory end-user. Perhaps because of their training, factory purchasers often have the engineer's proclivity to choose the most sophisticated technology they can afford regardless of its suitability to Chinese factory conditions. This is especially true if they work for state enterprises which operate in protected, low-competition environments. State enterprises are rarely, therefore, big buyers of used equipment. Managers of smaller private factories, on the other hand, are much more conscious of cost/quality trade-offs and are much more aggressive in seeking low-cost production technologies, even if they are occasionally second-hand.

Price

I/E corporations, always important players in any government-funded purchase, are more concerned with the up-front price than with technology. Since information about past purchases is widely circulated among I/E corporations, the supplier's distributor can expect requests for reductions of 10-30% on the asking price, based on past purchasing contracts of similar equipment. At this point, the quality of the distributor's relationships and market intelligence is crucial; competent distributors have a good idea how much the ministry can afford to pay and what other non-price factors will help secure the order. If the gap between both parties gets below US$10,000 on equipment valued at US$250,000-US$500,000, these non-price factors become important, the most significant being after-sales service, training, and foreign travel.

After-sales service

There was a time in the late 70s, soon after the CCP had instituted its "open door policy", when I/E corporations had a policy of preventing direct contacts between foreign equipment suppliers and factory end-users. If imported machinery malfunctioned, everything was done to repair it in-house; if that failed, the machine was carted out of the factory and transported to a major city where a representative sent by the supplier would have access to it, often in the backyard of a hotel. Since then, Chinese attitudes have moved to the opposite extreme; as long as the warranty lasts, factory end-users have no hesitation to call for help at the slightest sign of problem. After-sale service has become a required component of any sales effort in China and an important way to develop loyalty and repeat business.

Training

Skilled workers are at a premium in Guangdong and turnover rates are high. Most workers only wish to return to their home villages with some savings after their two-year contracts expire. Some turnover is also structural; state enterprises send managers to Guangdong for two years of training. As a consequence, foreign exporters planning for long-term growth in China must view training as a form of technology transfer requiring sustained effort.

Foreign travel

Visits to the supplier's head office by parties of ten to twelve people are a common feature of sales contracts. It is easy to be cynical about these requests, but they do offer a good opportunity for suppliers that are new to the China market to reshape the brand-ranking mentality so common among PRC managers.

COMMUNICATION

So much has been said about the excruciatingly slow pace of Chinese bureaucracy and its tendency to move only when strings are pulled that many North American suppliers assume all Chinese businesses conduct their affairs that way. This often gives rise to communication problems when the needs of Guangdong customers are misunderstood. Two particularly frequent problems are slow responses and *guanxi*.

Slow responses

While the Chinese government is often slow, Guangdong enterprises and, particularly, those in Shenzhen, are used to Hong Kong's intense business style. When they send a fax or telex to a supplier in the US or Canada, they expect a full response the very next day or they will take their business elsewhere. All too often, North American firms fail to respond or do so incompletely and and too late.

Guanxi

Guanxi means "connections". To have *guanxi* is to have friends who can be counted on to do a favor and bend the rules on your behalf. It is a required asset to successfully do business in the PRC; however, it no longer has the selling power in southern China many foreigners ascribe to it. The competitive climate in Guangdong is such that very few managers can afford to buy second-rate equipment simply because a friend is making the sales pitch. That kind of "connection" simply no longer exists. The *guanxi* that remains effective is trust which takes time to build; abuse it, and you have lost a customer for life.

THE PRC CUSTOMER: CONSUMER PRODUCTS

Although there exists no single "China market" because of local protectionism, certain demographic and regulatory trends have given foreign exporters more opportunities than ever to market consumer products in the PRC. We wish to thank the Hong Kong Market Research firm, SRG China, for many of the statistics contained in this chapter.

A MIRAGE?

The allure

According to the Hong Kong survey research firm, SRG China, the PRC market annually consumes 60 million watches, 30 million cases of cigarettes, 30 million bicycles, 20 million TV sets, 15 million radios, 8 million washing machines, 6 million refrigerators, 6 million tons of beer, 3 million cameras, 1 million motorcycles and half a million motor vehicles. Such figures were, in large part, behind the rush of American companies to China in the early 1980s. And yet, most quickly discovered that high custom duties, laws requiring that joint ventures balance their foreign exchange receipts and expenditures, and domestic marketing barriers effectively cut them off from the consuming public. Has anything changed since then?

The reality

China does not encourage the import of consumer goods due to the shortage of foreign exchange. Certain demographic and economic developments during the 1970s have, however, dramatically increased the willingness and capacity of Chinese consumers to pay for foreign and joint venture products. Furthermore, channels are increasingly opening up to allow foreign exporters, assisted by Hong Kong distributors, to test their products in the PRC and develop a certain market share. Nevertheless, the fact remains that long-term penetration of the PRC's consumer markets (there is no single market due to local protectionism) is much easier and more secure if the foreign supplier eventually establishes a manufacturing joint venture in China.

THE DEMAND SIDE

Urbanization

In the late 1970s, the people's communes were disbanded and farmers were allowed to till their own plots. This "rural responsibility system" exposed the long-lasting issue of surplus labor latent within China's rural economy. Half of China's 400 million farmers were no longer required on the farm. This development, along with lower restrictions on the movement of peasants after 1984, prompted most people to move to towns where almost half of them (93 million) found work in small township enterprises. Others moved to the cities or became part of China's huge floating population. The numbers of towns rose from 2,176 in 1978 to 11,481 in 1988, an increase of 427%. More importantly, the percentage of China's population living in cities or towns rose from 20% in 1981 to nearly 50% in 1988. This has had momentous implications for consumer demand. Relocating to towns means entering a money economy, the adoption of city consumption patterns, a greater sensitivity to advertising, and explosive demand for consumer goods.

More income

According to SRG China, the number of households reporting incomes of more than Rmb 7,200 (US$1,332) a year has grown dramatically, especially in southern urban centers (see Fig. 4-4). Unreported, however, are the millions of private businessmen and managers of independent companies earning ten times that amount. At the upper end of the scale, it is estimated that the number of people in China with an average monthly income of HK$5,000 or more, the salary level of mid-level consumers in Hong Kong, has risen to 50 million or about eight Hong Kong-size markets. And while Guangdong is currently the most prosperous market with a population of 64 million, many analysts feel that the area of greatest

Figure 4-4

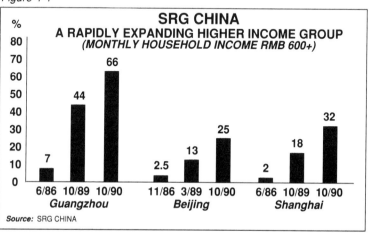

potential may be Shanghai and its satellite cities, a market of 150 million people.

This is cash income. Many urban enterprises also supplement this with income in kind, i.e., durable consumer goods and daily necessities which are supplied to employees free of charge or at nominal prices in an effort to attract and keep skilled workers. With income in kind now averaging 20% of urban incomes, institutional purchasing by employers of goods such as bedding, fabric goods, toothpaste, hair care products, food and other daily necessities is having a major impact on consumer awareness of new products. Toiletry brand choice is a particularly good example of the impact of workplace purchasing. Workers often take showers in public baths, especially in winter, because many homes have inadequate hot water supplies. As people become more image-conscious, these workers increasingly appreciate the novel toiletry products supplied to them by their employers. When questioned about how they first came to use a particular brand, many will therefore say it was given to them at their workplace.

Improved living conditions

Under Mao's leadership, China's economic policies chronically neglected its urban infrastructure, housing in particular. The building boom which followed upon his death was an integral part of the "marketization" of the Chinese economy that began in 1978 and continued throughout the 1980s. Increasingly, people are buying their own homes and are more willing to spend money to improve their living environment. This trend, along with the increase in living space per capita (see Fig. 4-5), is spurring demand for electrical appliances, quality construction materials, paint, furniture and household products.

Figure 4-5

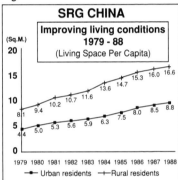

Hungry retailers

Amid this growth in consumer buying power, China's retailers have been forced to drop their "take it or leave it" attitude towards customer satisfaction. Firstly, the number of retail outlets increased seven times from 1.5 million in 1980 to approximately 11 million in 1990, from large department stores to streetside entrepreneurs selling Hong Kong-style products. This has generally intensified competition between retailers, especially for fast-moving and fad products since at least 50% of these outlets are now in private hands. Secondly, most major department stores have seen their state subsidies gradually cut in the name of decentralization; they are no longer allowed to be passive spouts at the end of the state production pipeline but must learn to react quickly to market forces. As a consequence, they have shifted to a contractual/bonus employment system to improve customer service; gone into direct sourcing and wholesaling to diversify their sources of income; and striven to secure the right to sell foreign or joint venture products known to be popular and whose prices are set by supply and demand.

Figure 4-6

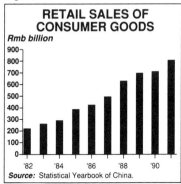

RETAIL SALES OF CONSUMER GOODS
Rmb billion

Source: Statistical Yearbook of China.

THE SUPPLY SIDE

Freer joint ventures
From 1984-1989, retail sales in China had an average annual growth rate of 18.6%. In the 1988-1990 period between the two waves of

reform, however, the growth in consumer spending slowed to a halt because of the government's anti-inflation austerity policy. Since then, retail sales have picked up again (see Fig. 4-6). With individual savings ballooned to nearly Rmb 1.2 trillion (US$224 billion) in 1992 and consumers refusing to buy many of the shoddy products of state enterprises (of the 20 million PRC-produced TV sets in 1991, half were unsellable), the central authorities have decided to encourage spending to forestall the inflationary havoc pent up demand could wreak on the economy. One way to do this has been to allow foreign joint ventures to sell a larger percentage of their production to the domestic market, regardless of what percentages were agreed to in their original joint venture agreements.

Porous borders
The easy convertibility of the Rmb in Guangdong Province (see Reforms under Deng), corruption among customs officials and smuggling have led to the growth in southern China of a huge underground entrepot trade in imported goods. This has made it possible for foreign exporters to test-market products or create a demand for products in the PRC without first going through the expense of setting up a joint venture. In most cases, however, these products must either be well-known brands or be perceived in the PRC as popular products in Hong Kong (see next chapter).

Retail joint ventures

A rush of property deals in 1992 by Hong Kong investors in China has highlighted two policy changes which will facilitate the foreign exporter's access to the Chinese consumer.

First, the central authorities are lifting restrictions on foreign participation in domestic retailing in the form of joint ventures. Hong Kong's Pacific Concord Group was the first to take advantage of this opening, seeking greater control over the marketing of its own joint venture products such as electronics, leather products, watches, garments, toys and cosmetics. By 1989, Pacific Concord had a network of 1,200 retail outlets in China. Then came Yaohan International, a Japanese retailer headquartered in Hong Kong, with a joint venture store in Shenzhen. Yaohan is perhaps more significant for foreign exporters because it has no manufacturing joint ventures in China and sells only imported goods. As more and more Hong Kong retailers follow in their footsteps, the transition from selling in Hong Kong only to selling in the PRC will be easier for products with a good track record in Hong Kong.

The second policy shift was to allow local governments to sell "land use rights", relatively clear 50-year leases on property, to both domestic and foreign buyers. For Hong Kong investors who are used to "buying" land from the Crown as 99-year leaseholds and who see retail and property investments as two sides of the same coin, this new policy has had the effect of making China's property market an extension of Hong Kong's own market.

Some examples of recent property deals as reported in the Far Eastern Economic Review (July 16, 1992) include:

- Sung Hung Kai Properties of Hong Kong signed a US$255 million contract with Beijing's Dong An Commercial Group to build a multi-storey shopping center on Beijing's busiest shopping street, Wangfujing.

- Hong Kong retailer Sincere and a Chinese partner will build a US$12.8 million department store on Shanghai's prestigious Nanjing Road.

- The Hong Kong subsidiary of Thai agribusiness conglomerate Charoen Pokphand signed a Letter of Intent to build a US$270 million shopping center in Guangzhou.

Flexible marketing

The collapse of the central government's subsidy system in the wake of reform has forced China's wholesalers and retailers to try new ways to increase revenues. The US Company, McCall, illustrates how the Chinese consumer can be reached if the foreign exporter finds a good distributor, proceeds one step at a time, and is willing to experiment with new distribution channels that respect the Chinese partners' needs.

[McCall] displays the virtues of hiring people who know how to work the Chinese system. Paul and Stacy Condrell met in 1983 when they were both studying Chinese in Beijing. After Mr. Condrell finished law school, they set out to become entrepreneurs in China.

They tried, with little success, to sell Burpoe Seed Co.'s premium vegetable seeds. Recognizing a need elsewhere, they sent a letter in 1986 to McCall explaining how China had 200 million sewing machines, plentiful fabric and several hundred million people eager to wear something other than Mao suits.

Since then, this energetic couple, both in their early 30s, has through trial and error created a unique all-Chinese sales system for McCall that last year sold 115,000 patterns through outlets in 50 stores in 40 cities. Sales are projected to reach 200,000 units in 1992.

Paul and Stacy, both fluent in Chinese, make sales calls to convince state stores that McCall patterns will stimulate fabric sales. They then install a McCall sales counter and bring in a trained clerk, paid by commission.

The patterns, imported from the US by a state trading company, are distributed by a cash-starved think tank, the Beijing Garment Research Institute, which the Condrells appointed as the company's exclusive agent in China. Each Chinese organization gets a small cut of the $2.40 sales price of each pattern. The Condrells and a few staff members nominally assigned to the Institute, however, do most of the work: it is basically a foreign sales operation in disguise.

The Condrells are masters of free publicity. Both national and local Chinese television stations and newspapers regularly run features that teach women how to use McCall patterns. The Condrells have two full-time publicity writers on staff. A Chinese magazine publisher, seeking a new magazine concept for a rapidly changing marketplace, has issued a McCall catalog in magazine form at no cost to the company. Mail orders are starting to boom.

"There are a lot of companies that could make it here if they put in our kind of operation," says Mrs. Condrell.

Her husband believes that the best rule for operating in China is to make your own rules, so long as they fit within the vague parameters that China sets for foreign and local businesses. "You can't get people here to say, "Yes, this is all legal" ", Mr. Condrell says. "Flirting with the edges of what is permissible is the way you have to do things in China. That is the way China works. You have to live with a certain amount of risk."

Source: Reprinted by permission of AWSJ (March 16, 1992).

PURCHASING BEHAVIOR

A youth market

The average age in Shenzhen is 28 while forty percent of Guangzhou's population is less than 25. Urban centers in southern China are, therefore, a growth market for baby, child and youth-oriented products. Milk powder, for example, is imported in large quantities because it is in short supply in every province except Heilongjiang in the north. Another expanding sector is the market for educational products and services. China's one-child policy has meant that parents lavish an important part of their revenue on the best clothes and foods for their "little emperors". In exchange, they insist on achievements from their offsprings which are almost exclusively defined in academic terms. Finally, young women are emerging as an important market for cosmetics and fashionable ready-to-wear apparel. The response of Guangdong women to the direct selling methods of Avon Cosmetics' have exceeded all expectations. Market research has identified the "typical" Avon customer as an urban resident between 20 and 35 years of age who uses cosmetics either on a regular or occasional basis but is willing to spend more on foreign products because of their greater quality.

Pockets of wealth

There is a big income gap between the segments of the population earning HK$5,000 or more per month – private entrepreneurs, actors, managers in foreign ventures, some officials in a position to squeeze foreign-currency earners, etc. – and the rest of the urban population. The former think that joint venture products are still not good enough and are willing to pay prestige prices for purely foreign imports, such as Benetton sweaters or Rolex watches. As for the latter, economically weak does not mean commercially insignificant. Their collective purchasing of consumer durables, food and personal care products (see Fig. 4-7) is important even if their limited incomes restrict them to the low value, unsophisticated end of the market.

Brand conscious

Thirteen years of reform in China have made PRC consumers very aware of international brand-name products and trends, especially in the coastal regions. If it is fashionable in Hong Kong, people in the coastal regions know about it and gradually influence consumers in such inland regions as Sichuan or Beijing. This derivative form of brand consciousness manifests itself in seemingly definite, nearly unanimous views among PRC consumers about what product is "best" within any production category along with a ready willingness to switch products if Hong Kong trends change. As a result, the life cycle of consumer goods in the PRC is growing shorter as new products are rapidly replaced by newer products which stimulate imports because domestic industrial production cannot always keep up with demand.

Figure 4-7

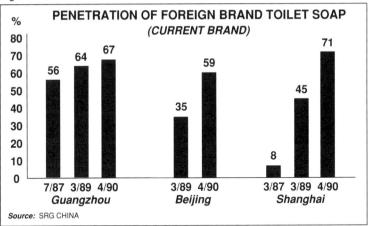

**PENETRATION OF FOREIGN BRAND TOILET SOAP
(CURRENT BRAND)**

Source: SRG CHINA

Sensitive to advertising

Since very few Guangdong consumers have the opportunity to leave China, much of their information about Hong Kong trends is derived from advertising on Hong Kong TV and radio programs. Indeed, advertising creates its own reality: certain exporters have boosted sales in the PRC by broadcasting ads in Hong Kong, giving the impression their products are more popular in the colony than they are in reality. Many PRC consumers are relatively unsophisticated and tend to accept advertising at face value.

Gift items

Gift giving to nurture relationships or at certain ritual occasions of the year like the Spring Festival is such an important part of Chinese life that imported gift products often span the income gap mentioned earlier. Gift items include a broad range of products that are small in size, consumed in public and expensive, and include such items as cigarettes, watches, liquor, chocolate, pens, and ties. To give a gift, especially an expensive foreign product with snob appeal, is to give "face" and show respect. Packaging, however, is very important. Many Hong Kong distributors add value to foreign gift items by repackaging them in boxes whose color and texture connote wealth and good fortune. These boxes have a value of their own and may be reused.

Different from Hong Kong

For all their fascination with Hong Kong trends, PRC consumers are not simply Hong Kong buyers with less money and urban polish. Geographically, they constitute a very heterogeneous cluster of markets where differences in customs, degree of affluence, dialect and level of modernization require tailor-made marketing strategies. Forty-three years of Communist rule has also left an institutional and behavioral legacy which is very foreign to Hong Kong. Realistic expectations, the ability to recognize business potential, mutual respect and the fortitude to weather the long haul are essential for commercial success in the PRC. By comparison, business life in Hong Kong is much more straightforward.

MARKET ENTRY AND THE HONG KONG MIDDLEMAN

*The PRC business environment poses several key challenges to foreign export-
ers of capital and consumer goods which require close cooperation between a
supplier and his distributor if they are to be surmounted. It is no coincidence
that many distributor agreements collapse around these same issues since they
often absorb more time and resources than the supplier originally anticipated.*

CAPITAL GOODS

In the marketing of capital goods,
disputes between suppliers and distributors about the required degree of
supplier participation usually center
around promotion, after-sales service,
and training.

Promotion

If we classify potential buyers
according to their source of funding
(see The PRC Customer: Capital
Goods), most foreign partners of
equity joint ventures or Hong Kong
manufacturers with assembly operations in Guangdong rely on suppliers
they came to depend on outside the
PRC. Promotion targeting both of
these groups is therefore done in
Hong Kong or wherever they are
headquartered and is relatively
straightforward.

There are **three avenues** used to
educate government-funded and self-
funded PRC buyers. The first and
least effective to contact are **PRC
companies with offices in Hong
Kong** in the hope that they wil act as
the supplier's agent in China. There

are over 4,000 PRC-backed companies in the colony and only 25% of
them bother to register their operations with the Hong Kong government or with Xinhua (the PRC's
"shadow government" in Hong Kong)
because they prefer to keep Beijing in
the dark about their business activities. PRC companies in Hong Kong
fall into four main categories: State
Council companies, such as China
International Trust & Investment
Corp (CITIC), ministry-level enterprises, such as MOFERT's China
Resources, companies representing
provincial governments, and enterprises representing city, county and
even township governments. The latter two include most PRC companies
in Hong Kong; over 1,000 represent
various groups within Guangdong
Province alone. At best, they represent the interests of their mother
organizations; at worst, they are busy
laundering state-allocated foreign
exchange by investing it in Hong
Kong's stock and property markets.
In no way are they willing or able to
serve as a foreign supplier's distributor. Suppliers are, therefore, received
with polite indifference and very little

of the information left with them reaches its intended audience in the PRC.

Trade fairs in the PRC are a second avenue and a good way to reach potential factory end-users. However, so many exhibitions have been organized in recent years that many of them conflict with each other, wasting the time and money of foreign and domestic participants. An exporter and his distributor must first sit down and agree on common goals and the resources required to reach these goals. What region of China are they targeting? Should they focus on larger national fairs or the more specialized mini-fairs? If the supplier is new to the PRC, how does he establish that all-important PRC track record without which large sales are impossible? Should he sell some demonstration equipment at a discount at the end of certain fairs? And if he does and the discounted price becomes public knowledge within the I/E corporations, how will he justify a more normal price when promoting the same equipment to other potential PRC buyers?

A third avenue is to conduct **technical seminars** in China. The time and place must be carefully planned and agreed to by both the supplier and his distributor because technical seminars can be a gruelling exercise. Typically, they attract a more diverse crowd than trade fairs, including factory end-users, ministry officials, I/E corporation reps, engineers from various research institutes, etc. They also cost more in terms of interpretation services, product literature, and hotel bills and offer a great opportunity for everyone to pick the supplier's brains for technical and proprietary information unless he is accompanied by someone who knows how to guide the presentation. Technical seminars do not sell equipment on the spot; however, they can influence important decision-makers if the supplier is assisted by a distributor who knows what he is doing.

After-sales service

After-sales service plays a crucial part in establishing a good track record in the PRC. However, PRC maintenance practices often give rise to friction between suppliers and their distributors because of the hidden costs involved. A poor response to these requests by the distributor will quickly ruin the supplier's reputation in the PRC.

In most PRC enterprises, preventive maintenance is still a rarity. Even though imported machines are often off limits to all but the best workers, they are generally kept running until they breakdown and only then will the distributor be contacted.

Within the warranty period, distributors can expect a high number of requests for assistance even for trivial problems which were dealt with by the training program offered to the end-user. Requests for help often disappear as soon as the warranty period expires. Distributors find, however, that some level of service must be maintained if their reputation and that of the supplier is to remain sound.

Post-warranty servicing must therefore be taken into account both by distributors and suppliers before sales are made.

Depending on the equipment involved, spare parts are rarely a big source of revenue for suppliers selling to state-funded buyers. For one thing, the budgets of state enterprise frequently make insufficient allowances for the hard currency needed to buy imported spare parts. Moreover, the PRC's trade liberalization measures designed to facilitate its entrance into GATT are giving factory end users access to spare parts from a variety of sources. No longer is the sale of equipment a guarantee that steady orders for spare parts will follow.

A basic source of conflict between distributors and suppliers comes from their different perspectives on the kind of partnership required to succeed in the PRC. Because mutual trust is a very important ingredient of business life in the PRC, many distributors try hard to achieve a uniform level of dependable service regardless of which supplier is involved in any given purchase. From their point of view, the closest approximation to the ideal supplier are certain Japanese or German firms who take a proactive approach to after-sales service because exports are a crucial part of their business. Instead of appearing only when problems arise, they visit their distributor and end-users regularly to get feedback and glean valuable market information. They view their relationships with a distributor as a close partnership allowing them to penetrate the PRC market further and faster than they could on their own. Their business style is intensely relationship-centered. North American suppliers, on the other hand, have a deal-centered business ethos and view the supplier-distributor relationship as a division of labor which allows them to forget about the PRC while they concentrate on their domestic markets. In many cases, exporting is treated as a stop-gap measure designed to pick up slack when domestic demand is weak. They tend therefore to hustle spare parts and provide a minimum of technical support, an approach which does not always fare well with Hong Kong distributors and PRC end-users.

Training

Training also often highlights the "deal" versus "relationship" orientation of suppliers and distributors. Because of low literacy levels and high turnover rates (see The PRC Customer: Capital Goods), training often must be seen as an ongoing function not as a one-shot deal. Both from the PRC end-user's perspective and that of his up-line ministry, machines embody technology and, thus, equipment suppliers are involved in a form of technology transfer, not just sales. This gives rise to expectations and costs which must be planned for and apportioned between suppliers and their distributors.

CONSUMER PRODUCTS

Compared to capital goods, the marketing of consumer products in the PRC confronts distributors and suppliers with a greater variety of possible market segments, distribution venues and promotion strategies, each with its inherent advantages and disadvantages depending on the supplier's goals and his willingness to commit resources to them. We will focus our attention on four basic issues where it is particularly important that suppliers and distributors be on the same wavelength: goal setting, channel selection (for the foreign currency and Rmb markets), promotion, and joint venturing.

Goal setting

Before the distributor selection process gets seriously under way, a foreign exporter must conduct enough research about the real market situation in the PRC to know **why** he is in Hong Kong and not in some other market; **what sales targets** he is aiming for in the short and long term; and **how many resources** he is willing to devote to reach these goals. To search for a distributor without a clear answer to these questions will start a bidding contest among candidates, none of whom will take their promises seriously. The result can only be disillusionment for the supplier a few months down the line.

Once a distributor has been selected, goal-setting must be carried out as a joint process, especially during the crucial first six months when signs of success are important to maintain distributor motivation. All too often, suppliers set unrealistic goals on their own, only to see their distributors order less than expected, challenge prices and question the competitiveness of their product line, all sure signs of sagging enthusiasm.

Channel selection

1. The hard currency (or FEC) market For certain product categories, such as gift items and luxury consumer products, distributors often concentrate their efforts on PRC state-controlled retailers which accept only foreign currency (e.g., duty-free shops, friendship stores, hotel shops, cross-border kiosks and certain joint-venture stores). Choosing such venues not only reaches a wide-range of customers with foreign currency – independent entrepreneurs, foreigners, joint-venture managers, households with relatives abroad, etc. – but it can also enhance brand recognition in the minds of the many shoppers who only come to browse.

To reach these stores, distributors sell to certain PRC companies based in Hong Kong: national I/E corporations, the China National Duty-Free Merchandise Corporation, and certain companies representing the SEZs. Because these buyers are very bureaucratic and the distributor loses control once the goods have been sold to them, it is a misnomer to say that anyone "manages" this channel. The degree of control on the Chinese side is, in fact, weakening since certain buyers, the SEZs in particular, are illegally reselling an increasing percentage of their Hong Kong purchases on the Rmb market.

Note: The hard currency market is sometimes referred to as the FEC market. Chinese currency is denominated in yuan and is available in two forms. One is the renminbi (Rmb), literally "the people's money", which is not convertible; the other type, the foreign exchange certificate (FEC), is convertible. The two currencies are theoretically at par but the FEC trades at a premium on the black market. It is worth noting that the ready availability of Hong Kong dollars in Guangdong has made it, not the FEC, the convertible "Chinese" currency of choice in that province.

2. The Rmb market This is the Holy Grail for most exporters of consumer products targeting the PRC. The channels used to penetrate this market can roughly be classified as underground or above board; many distributors use a mixture of both.

The **underground option** is open to distributors of foreign brand name consumer products already popular in Hong Kong, or perceived to be so by Guangdong consumers because the supplier has chosen to advertise his product heavily on Hong Kong television and radio stations. Such distributors are frequently contacted by brokers willing to pay with hard currency for shipping containers of their products. If the transaction goes through, these containers are either smuggled into China or shipped openly to small ports where customs officials connected with the broker can be counted on to levy import duties at very low rates. Because customs officers are sometimes rotated, a given broker might ship through Huizhou one week and then move on to Dongguan the next. Once the containers have made it into Guangdong, they are resold to retailers or to middlemen working in various underground "collection centers" many of which are in Shenzhen or Dongguan (see map below). These centers play an entrepot role, dividing incoming products into batches and distributing them to various regions in the PRC where they will be sold for Rmb. Profits of 200% or more are common for brokers but they must share their revenue with their network. These networks are so exclusive that a broker who is excellent for one prefecture of Guangdong may not be as effective in another. For this reason, Hong Kong distributors using this channel never grant exclusivity for a single province, let alone all of China; expertise and networks are too location-specific.

The main advantage of this channel from the supplier's point of view is that he is paid up front in hard currency and can test-market his product in the PRC without having to negotiate from a position of weakness with mainland wholesalers. Indeed, if his product proves popular, he can consider going "above board" and is then far better placed to ask for concessions from PRC wholesalers and retailers eager to get in on a sure

Pearl River Delta Economic Zone

thing. On the negative side, the supplier has no control over the destination, pricing and conditions under which his product is sold. For all he knows, this reputation may be tarnished because a high percentage of his products are reaching consumers in damaged or diluted form. A change in customs regulation could also kill his business overnight. Finally, he has very little recourse if copies of his product start appearing in the PRC selling at half his price. The results can only be the loss of his brokers unless he has new products to propose.

From the distributor's point of view, especially that of transnational distributors selling thousands of different products in the PRC, the underground channels have the additional disadvantage of endangering their good relations with the Chinese government, a vital ingredient for any firm whose main assets in the PRC are its reputation and network of contacts.

Distributors adopting the **above-board option** piggyback on the PRC's increasingly decentralized wholesale and retail systems in an attempt to forge new channels for imported and joint venture products. Because China has no national retail organizations such as retail chains and because the reform process is not progressing at the same pace in all provinces, distribution "solutions" arrived at in one region or city are not necessarily transferable elsewhere.

At the wholesale level, the PRC's three-tier system (the national and regional level, the provincial level, and the townships level) is being restructured to adhere to the principle of using major cities in economic regions as centers of distribution. Each city sets up wholesale agencies which it manages itself. Department stores are also allowed to operate wholesale businesses according to their needs. This has had the effect of freeing up wholesalers and retailers eager to sell new products for which there is strong demand. It has also, however, strengthened the hand of local governments wishing to protect "their" publicly-owned enterprises (whose products may sell poorly) because local jobs are on the line. Retailers may, therefore, be under pressure to refuse certain products or may work under a fixed ratio system requiring them to sell a certain number of local products for every product manufactured outside the province. Local protectionist measures of this kind target not only foreign imports but also PRC manufacturers from other, more competitive provinces.

To increase control, Hong Kong distributors choose at least two wholesalers per province, often wholesalers reporting to two different ministries, such as the Ministry of Commerce and the Ministry of Light Industry (remember the lack of "horizontal" communication between organizations working under different ministries. See The Maoist Legacy). Although it is the wholesaler's job to monitor and motivate his down-line retailers, the Hong Kong distributor can go a long way towards making his wholesalers more effective by lending his logistical and marketing expertise. Distributors commonly lend vehicles to reduce transportation bottlenecks as well as organize promotion packages, including seminars, display contests, sale contests, point of sale promotions, free promotional items, sports and song competitions, etc., all of which can be effective tools to create brand awareness and boost sales.

Promotion

Although they are very rarely encountered in their "pure form", it is conceptually useful to classify promotional strategies used by Hong Kong distributors in the PRC under two headings: pull and push strategies.

At the macro-level for all of China, an example of the **pull strategy** are the many companies relying on small distributors or trading houses (transnational distributors may be unwilling) to build up demand within the PRC by exercising the underground option. Only when a product has an established track record will the supplier consider switching to more formal "above board" marketing channels. At the micro-level with individual retailers, pull strategies are often used by distributors of prestige products or products, such as peanut butter and jeans, where demand is expected to grow slowly as Chinese customers learn to appreciate them. Cosmetics firms, such as Christian Dior, are willing to go one step at a time because image is very important to the consumers they are targeting. Typically, they will open one boutique in a high-class section of a major city, perhaps in a good hotel; cosmetics classes might subsequently be organized for certain clients. Only when demand has grown sufficiently will a second boutique be opened.

Avon's use of door-to-door sales girls in Guangzhou is a good example of the **push strategy** at the macrolevel (see the case study at the end of this chapter). At the micro-level, Hong Kong distributors using a push strategy often encourage PRC wholesalers and retailers to overstock to maintain motivation. They may also offer promotional packages including free samples, door prizes, and display competitions, etc. Push strategies generally cost more than pull strategies, require more marketing know-how from the distributor, and are used by suppliers of established fast-moving consumer products. Indeed, for products with a short shelf life, such as chocolates, the supplier's only option might be a push strategy if product quality is to be maintained.

Joint venturing

As was mentioned earlier (see The PRC Customer: Consumer Products), trade can only be a prelude to local manufacturing if the supplier is determined to secure a long-term position in the PRC's domestic market. By itself, trade leaves the supplier at the mercy of shifting customs regulations which can ruin overnight years of painstaking effort developing distribution channels. Certain suppliers will therefore follow a two-track system: trading with China while, at the same time, going through the process of negotiating for a joint venture or a wholly foreign-owned enterprise. If that is what the supplier intends to do, the distributor will want to know where he fits in this picture. Is he destined to disappear? And if so, why should he invest all the resources required to create distribution channels in the PRC only to loose everything three to five years down the line? It may be difficult to address this initially when the supplier is unsure whether his product has a market in China; however, this question is sure to arise as soon as he encounters signs of success.

The supplier has several options if his long-term goal is to set up a joint venture: he could eventually dismiss his Hong Kong distributor with some form of compensation and manage his distribution channels thereafter. Or, he could concentrate on production and leave the physical distribution to his distributor. Another possible option, if the distributor is willing and able, is to include him as a joint venture partner. Joint venturing has become a strategic option for certain distributors wishing to secure long-term relationships with certain suppliers (see Types of Distributors). Either way, cooperation between a supplier and his distributor can quickly deteriorate if the distributor is left unsure as to his long-term relationship with his supplier.

AVON'S FINISHING SCHOOL

Guangdong women have responded beyond all expectations to Avon Cosmetics' direct selling technique, both as customers and sales ladies. After the first year's door-knocking, sales are double the projected figure, according to Barry Wong, general manager of the US company's local venture.

"It is definitely better than we expected," he said. "The management is very excited about business here. It looks to be booming."

Avon Cosmetics, the first company to use the direct selling method in China, has had no trouble recruiting its door-to-door sales girls. More than 6,000 have signed up and been through its training course.

Women have been eager to obtain the Avon training and earn possibly double their normal salaries. Earning 20-25% commission on sales, those who work hard can expect to take home about Rmb 700-800 a month – at least double their local salary. The record monthly sales figure for one woman topped Rmb 10,000, earning her about Rmb 2,500.

"As many as 95% of women will open the door. Most people really sit down and listen," said Wong.

Avon secured its foot in the China market by setting up a joint venture with Guangzhou Cosmetics Factory (GCF) with an initial investment of US$1 million. Avon owns 60% of the venture, GCF 35% and a Hong Kong partner 5%.

Ingredients for its cosmetics range are imported and manufactured in GCF's Canton factory which has been upgraded to Avon standards. "At the moment, there are no local ingredients that we can use," said Wong. "Maybe there will be in the future."

Sales are coordinated through two Avon centres in Canton and one in Foshan. A new center is due to open in Zhongshan. Within three years, Wong hopes to have the entire province covered. He said that the joint venture manages to export enough to other countries in Asia to balance its foreign exchange needs, though it boosts its hard currency by purchasing other local products for export.

The Canton factory produces more than 120 different products in the Avon range, which is soon to be increased. Retail prices are cheaper than imported products but more expensive than local. "Only a small percentage of people say that our prices are too expensive," said Wong. "The majority of young women here would rather spend their money on imported or joint-venture products [than local ones]."

"As average incomes increase, people are looking for quality. We are putting our quality and service on the front

line," he said. Avon offered a wider range of skin-care products for different skin types and a wider color choice for cosmetics than any local company. It faced little competition.

"Local companies have had to upgrade their products in response to the joint venture, but they don't have the hard currency to buy from outside to significantly improve their ingredients," said Wong.

In being taught how to make the best of their appearances and to be assertive in their contacts with people, women in Canton are being attracted to Avon as the next best thing to a finishing school. Before they go out on the road, they attend makeup, skin-care and product knowledge courses.

Secretaries, students, shop workers and private businesswomen are among those who have come forward to be Avon ladies. About 70% are in the 18-35 age range, though there are also women in their 40s and 50s.

"At the beginning, Avon ladies were a bit embarrassed to approach strangers," said Wong. "But we train them how to do this. They welcome the earning opportunity and really enjoy being Avon ladies. They say that before they joined Avon they had never had a chance to test their own abilities regarding sales and self-confidence."

"At first we were concerned about them running the business side. But there has been no problem. We give them credit but most come in and pay cash for the product."

Because the cosmetic products are unavailable through retail outlets, knowledge of the Avon range is spread through a heavy advertising campaign, with an emphasis on television commercials. All sales materials are printed locally and next year Wong hopes to shoot a commercial in China.

Source: *China Trade Report*, November 1991.

PREPARING
THE GROUNDWORK

SOME FUNDAMENTALS

Many sales representatives stumble in their search for good Hong Kong intermediaries because distribution selection is considered to be a simple matter requiring a few days at the most. This is a good recipe for disaster.

TWO APPROACHES

Advice on how best to approach supplier-distributor relationships is frequently contradictory. Depending on who is speaking, exporters may be told that cultural norms in Hong Kong make long-term relationships based on trust a precondition of success. Other experts, however, believe that distributors, like used car salesmen, have a vested interest in starving their suppliers of information. What, then, is the attitude exporters should adopt when evaluating prospective distributors or agents?

To clarify the issues involved, we can divide advice of this kind into two theoretically opposed, yet practically compatible, schools of thought: partnership versus carrot & stick.

The partnership school

Distributors and academic theoreticians of competitiveness are strong partisans of this approach. It encourages suppliers to aim for very close and long-term alliances with the distributors they select based on common goals and complementary strengths. They therefore deny any fundamental conflicts of interest between suppliers and distributors and point to the benefits of trust, early admission of mutual vulnerability, a

win-win approach to negotiation, an open exchange of information, and active sharing of resources by both partners at every level (technology, personnel, promotion and advertising, etc.). Some will also add a cultural dimension to this line of reasoning by stressing that Chinese distributors have a cultural preference for long-term, highly personalized relationships.

The carrot & stick school

The opposite view holds that distributors should be kept at arm's length because their interests are, in some ways, fundamentally at odds with those of their suppliers.

- Distributors have a vested interest in limiting their suppliers' access to information. Information, after all, is ammunition the supplier can use to criticize his distributor's performance. Moreover, the less a supplier knows about the market, the more indispensable is the distributor.

- From the distributor's perspective, he is caught in a black widow relationship with his supplier: not only can he lose the account if he does poorly, he can also lose it if he succeeds too well and makes it worth the supplier's while to set up his own direct sales operation. The best defence therefore consists in stretching out his useful life by doing the bare minimum to keep the supplier happy and no more!

- From the supplier's perspective, his main competitors may not be the Australian or Japanese products similar to his own being sold in Hong Kong, but the many unrelated products being carried by his distributor. It is them, not the Japanese, who are limiting his share of distributor resources. One of his key tasks, therefore, is to motivate his distributor's sales team to devote an inordinate amount of time pushing *his* products as opposed to those coming from other suppliers. As long as this does not negatively affect the overall profitability of the distributor's company, senior management will most likely not object.

These and other conflicts of interest have led many successful exporters to apply a carrot & stick approach to distribution. The first rule initially is to be wary, not trusting. They accept the fact that information and opinions provided by distributors must be largely discounted because they are incomplete and self-serving. They know that there is a basic conflict between their goal of rapidly increasing sales and the distributor's desire to keep them dependent as long as possible. Above all, these exporters believe that above-average performance from distributors does not happen spontaneously because both parties are "partners". It depends on performance-based, *quid pro quo* incentives for every level of the distribution team, and active courting by the supplier to ensure that his products absorb more than their share of distributor resources. Hence the term "carrot & stick".

Our approach

The partnership model comes closest to the truth when suppliers and distributor are both large, roughly equal in size, and the product being sold is complex enough to require a lot of pre-sale and after-sale coordination between both parties. The sale of capital equipment in the PRC is a good example (see Market Entry and the Hong Kong Middleman).

Our research indicates, however, that for most exporters, the carrot & stick approach is far more realistic, especially for small and medium enterprises new to Hong Kong. Unless their products are so innovative as to offer extraordinary growth potential, the large distributors will probably not be interested because the funds required to build up sales to profitable levels *for them* cannot be justified by the eventual returns. This means that small, hungry and (perhaps) naive players will almost always be better in the early days to painstakingly develop a market for new products because small distributors must get out and sell in order to survive. At the same time, they require more supervision and support by their suppliers, and respond very well to the carrot & stick approach.

This is, therefore, the basic point of view we have adopted in this book.

The fact that Hong Kong is culturally Chinese changes nothing in our approach. Indeed, as long as contracts are respected, Hong Kong businesses are used to frequent switching between "partners" (see Management Hong Kong-Style).

THE PERILS OF GOING BLIND

Most sales efforts in Hong Kong amount to very little because many sales reps have no plan, little time and insufficient sales backup, the result of the low priority North American manufacturers have traditionally given to exports.

No plan

Sales reps targeting the Asia Pacific region frequently receive only ambivalent support from their superiors. The attitude of many senior managers is: "Bring me an interesting offer and I will approve it." This leaves the sales rep with no basis for a plan. As a consequence, Hong Kong end-users and distributors have grown used to a daily parade of visitors on a fishing expedition, hoping something will fall in their lap, never to be heard of again. With no preliminary sales promotion plan, inadequate background information on the Hong Kong and PRC markets, and no idea as to which distribution mechanism is more likely to bring the best overall results, these representatives lack the basic tools required to attract and select good distributors.

A more useful approach would be to say: "I want to get into China; I have set aside X dollars to do it. Find out how." This gives the sales rep sufficient resources to effectively plan his distribution strategy.

No time

Ambivalent support from senior management also creates pressure for quick pay-backs, the result being a shotgun approach to distributor selection which can be very costly in the long run (see Weak Points and Trouble Spots). For lack of time, one finds sales reps accepting at face value data fed to them by distributors; initiating negotiations with distributor companies at trade shows without first completing any independent research in the marketplace; and seeking advice from the uninformed while junketing with trade missions. The classic mistake is the whirlwind tour of Asia by marketing managers who spend a few days in each of the main capitals, talk to two or three potential distributors, and make a decision before leaving town, often granting exclusive rights to the candidate who places the largest initial order.

Inadequate sales back-up

Some manufacturers succeed in locating a competent distributor only to see their relationship slowly go sour because their export sales departments are too disorganized to respond adequately to the distributor's requests. The best efforts to identify, screen and select good distributors will likely be in vain if responsibilities for certain vital functions are not clearly defined and placed in capable hands. These key functions are:

Management: An experienced export manager is a must.

Sales Coordination: Export planning, labelling and packaging, development of promotional material, etc.

Accounting: Maintain timely product cost estimates; deal with L/Cs and other trade-related bank instruments; advise management about expenses, new product launches, etc.

Transportation: Negotiate, schedule and supervise land, sea and air transportation.

Field sales liaison: Keep direct contact with field reps and distributors.

This may seem trivial, but one of the main complaints Hong Kong agents and distributors have about North American suppliers is the number of messages they send which somehow fall through the cracks. Suppliers who do this once too often lose the battle for distributor attention. No amount of correspondence after the fact will change this.

GETTING THE BIG PICTURE

Before evaluating distributor candidates, some crucial preliminary steps are required: acquire a basic knowledge of Hong Kong statistics as they relate to your product; study your competitors' products and competitive position; and investigate your target customers' buying habits. Failing to follow these steps reduces the selection process to a game of chance and greatly weakens your bargaining power when negotiating distribution agreements.

PRE-TRIP MARKET ANALYSIS

For Hong Kong

If you wish to know how much your country exports product X worldwide, your best source of trade statistics is your own government. If, however, you want to know how much of product X Hong Kong imports from all sources and how much is locally manufactured, contact the nearest office of the Hong Kong Trade Development Council (HKTDC or more simply TDC) to obtain the relevant information (see Annex 2).

It is important to ascertain the classification code for your product because vague requests for information cannot be answered precisely. As of January 1, 1992, Hong Kong's official trade statistics are classified according to the **Harmonized Commodity Description and Coding System** or **Harmonized System**, (HS) for short. Requests for information using the older **Standard International Trade Classification** (SITC) system will be answered but traders

dealing regularly with Hong Kong should purchase a copy of the *Hong Kong Imports and Exports Classification List (Harmonized System), 1992 Edition*. It costs HK$43 and is available from HKTDC or the Hong Kong Census and Statistics Department (see Annex 1).

All of HKTDC's offices worldwide (see Annex 2) have access to an on-line trade information system called TDC-Link. The TDC-Link information menu includes: updates on the 10,000 overseas buyers who visit HKTDC each month for supplier introduction, classified lists of overseas importers (180,000) and Hong Kong manufacturers and traders (48,000 companies), textile quota information, shipping schedules, trade statistics, import regulations, industry and market profiles, trade event news, HKTDC news and promotion schedules, and FOREX and interest rates. Although much of this information is organized for the convenience of foreign importers of Hong Kong products, a foreign exporter can easily gain access to the

GLOSSARY OF HONG KONG TRADE

Transshipments *are goods which are consigned directly from the exporting country to a buyer in the importing country even though they may be transported through or stored in Hong Kong. Because they are in transit and do not clear customs, transshipped goods are not included in Hong Kong's trade statistics.*

Imports *include all goods imported into Hong Kong for domestic consumption or for subsequent re-export.*

Domestic exports *are exported goods which are either naturally produced in Hong Kong or the result of a manufacturing process which has permanently changed the shape, nature, form or utility of imported raw materials. Processes such as diluting, packing, bottling, drying, assembly, sorting, decorating, etc., do not transform imported raw materials sufficiently to qualify as Hong Kong origin.*

Re-exports *are products originally imported into Hong Kong before being exported which have not undergone sufficient transformation while in Hong Kong to qualify them as domestic exports.*

Retained imports *are defined by the Hong Kong government as the difference between total imports and re-exports. This understates the value of retained imports because the re-export margin (the costs of transportation, insurance, storage, purchasing and minor processing as well as the profits made by the Hong Kong intermediary) is not considered. The HKTDC has estimated this re-export markup to be approximately 15%. The re-export markup for PRC goods, many of which are the products of Hong Kong outward processing activities in Guangdong, is considerably higher at 25%.*

import statistics which relate to his product. The most basic include:

- Hong Kong's imports of your product category during the last three or four years.
- Hong Kong's re-exports of the same product category during the same time period.

Subtracting re-export figures from imports for each year will leave you with that part of imports actually consumed in Hong Kong: its retained imports. Because Hong Kong's re-export figures include both the original cost of re-exported products and a 15% margin for the middlemen (see the glossary opposite), a more accurate retained export figure can be obtained if that 15% is removed from the re-export figure [i.e., re-exports − (15% x re-exports)]. We are therefore left with a basic formula:

Retained imports = imports − [re-exports − (15% x re-exports)]

Applying this formula for the last three or four years should tell you whether the market is growing and, if so, at what rate.

EXAMPLE:

Furniture and parts thereof (includes many smaller categories):

HS code: 9401

SITC code: 821

        Hong Kong imports (1991):   HK$ 4,384 million

                exports (1991):   HK$ 2,181 million

Retained imports (1991)   = $ 4,384 − [$ 2,181 − ($ 2,181 x 15%)] *

                      = $ 4,384 − $ 1,853.85

                      = HK$ 2,530.15 million

    *If most of this furniture is imported from the PRC, the re-export mark-up would have to be closer to 25%.

To get a clearer picture of who your competitors are, more data can be requested from TDC-Link:

- Hong Kong's imports of your product category by country of origin during the last three or four years.

- Hong Kong's re-exports of the same product category during the same period by countries of origin.

Using the same calculation method described above should give you a good sense of import market shares within your product category and whether these shares have changed over the last few years. For names of actual companies, HKTDC is not always the best source. To find out which German companies export products similar to your own to Hong Kong, you may have to contact the German government or go to Hong Kong. TDC-Link can nevertheless inform you which Hong Kong trading houses import your product category, simplifying the research process once you arrive in Hong Kong.

For the PRC

Unless you possess a highly developed export department, it is very difficult to determine from a distance the amount a given province imports of your product category. Aggregate numbers for the PRC as a whole are available but their accuracy has often been called into question (see The Birth of Greater China). The best sources of information, especially on Guangdong imports, are the Hong Kong distributors and trade companies that deal regularly with the PRC. For general information on provincial economies, TDC-Link can supply you with two-page overviews which are regularly kept up-to-date.

Tariffs, regulations, etc.

A comprehensive and easy-to-use source of information on tariff regulations, health standards, labeling and packaging requirements for Hong Kong and the PRC is the regularly updated publication, *Exporter's Encyclopedia*, by Dun's Marketing Services. Most trade libraries have a copy. Another good source of information on import regulations are the Hong Kong government's many representative offices around the world (see Annex 9).

Warning

In the early stages, trade statistics and market surveys can be very useful when prioritizing your export targets. They can also be useful when learning how a given country's economy is structured and how its cultural traditions, consumer habits and technical maturity might affect your product. However, surveys, newsletters and macro-economic reports frequently contradict each other when you narrow your focus to a single topic. Worse still, they can engender a form of paralysis by providing you with a wealth of intelligent-sounding excuses for doing nothing.

If the trade statistics seem to indicate that there is a worthwhile market for your product in Hong Kong or in the PRC, the next step is to go there for a visit. The real data – the solid market information – can only come from talking to people with experience. This does not deny the value of statistical data; indeed, a good understanding of Hong Kong's trade statistics as they relate to your product will give you the self-confidence to talk to those who really know what is going on. As your experience broadens, statistics never lose their importance. In fact, they are a useful tool when goading a distributor or when arguing over sales targets and marketing strategies. You will never, however, reach that stage if you stay home reading reports.

PROSPECTING IN HONG KONG

What to bring

Here are some things to bring along with you:

- Business cards.
- Annual reports and/or one-page company descriptions (see Networking and Negotiations).
- Product brochures and specifications.
- Retail (not net) price lists.
- Company letterhead stationery.
- A preliminary list of contacts along with copies of any correspondence with them.
- A preliminary list of interview questionnaires.

Goals

Your main goals, at this point, are twofold:

- To know the competition: Who exports to Hong Kong products similar to your own? Who distributes them and through what channels? What is the pricing structure, etc?
- To define the customer: who, what, when, where, how and why do they buy?

These goals may seem daunting; but Hong Kong is perhaps the easiest place in Asia to gain introductions and gather advice. As long as you know what you want, present yourself well, and stress the potential advantages to your listener, achieving these goals should not take longer than a couple of weeks, considerably less if you are experienced in this (see Networking and Negotiations).

Language

In most cases, language will not be a problem when dealing with the institutions listed below because they have learned to function in English. Interviewing end-users, on the other hand, sometimes requires the services of an interpreter. The best way to minimize this added expense is to gather as much information as possible from English speakers before approaching other groups; limit your interviews with Cantonese speakers to a few key points; and concentrate your Cantonese interviews in one or two days. Whatever happens, don't let language stand in the way of getting the information you need.

Government

The first stop for many is to visit their country's trade commission in Hong Kong (see Annex 10). In most cases, this is a very wise step. Foreign trade commissions advise thousands of business visitors every year and have enough clout to open many doors. In some cases, that distribution is the responsibility of the local staff because of their knowledge of Cantonese and local business practices. Make a point of befriending these local officers; their advice can be very valuable.

HKTDC

Reference has already been made to TDC-Link. Their trade offices and library are worth visiting for information on competing products in Hong Kong (see Annex 2).

The Hongkong Bank

Another good source of contacts and trade information is the Hongkong Bank's Trade and Credit Information Department (Level 9, 1 Queen's Road Central, Hong Kong Island, Tel: 822-3535 or 523-8422, Fax: 868-1646). Availing yourself of their help does not require that you have an account at the Hongkong Bank; however, a letter of introduction from your own bank or trust company describing the nature of your company will help you get more attention.

Chambers of commerce

Whether or not the individuals heading these organizations (see Annex 8) have expertise in matters of distribution depends a lot on the person involved. Chambers are, first and foremost, networking forums designed to inform and defend their members by exchanging data and acting as industry-watchers. Used with care, they can be a good source of contacts.

Law and accounting firms

Another good source of contacts and advice (see Annexes 5 and 6) are law and accounting companies, especially if you have a letter of introduction from your own law or accounting firm. Accountants generally have a larger client base and a deeper understanding of their clients' business activities than lawyers. Pay most attention to those lawyers and accountants who do the deal-making for the firm: they are the best-informed sources.

Industry and trade associations

Like the chambers, these are basically networking forums. They can, however, help you meet retailers and manufacturers who can greatly broaden your perspective on customer buying habits, industry trends, and distribution mechanisms (see Annex 8).

Transnational distributors

As long as you steer away from any negotiations and cross-check what you are told, large distributors (see Annex 3) can be a very useful source of data and advice, particularly if your product is compatible with the ones they already carry. Some may even be willing to allow you to accompany their salesmen for a day, a very valuable opportunity for you to question them on the nature of the work, market trends and the internal division of labor at the firm. Small distributors should be avoided at this stage because they have no time to waste offering free information, and will certainly not give you access to the sales force for fear that you might establish a direct, selling relationship with potential buyers.

Market research firms

These are a good source of off-the-record information, especially on consumer products and trends (see Annex 4).

Academia

Universities and technical institutes can often be an excellent source of information in the field of capital goods and occasionally consumer products. Some academics may also know a great deal about the PRC.

Warning

While group presentations with distributors might seem a great way to save time, the consensus among experts is that the disadvantages of this approach outweigh the advantages. The main shortcomings with this approach are the loss of confidentiality as price lists and discount policies enter the public domain; an overemphasis on your product's weak points as everyone in the room evaluates how you react to criticism; and reduced flexibility during negotiations. Individual interviews may require more time but they allow you to learn as much as the person being interviewed while limiting the negative effects of your inexperience.

DEFINING THE CUSTOMER

The only way to ensure that you remain in the driver's seat throughout the distributor selection process is to have a firm grasp of whom your target customers are before the negotiation process begins. Done well, it should leave you with a good idea of which distribution mechanism is best for you and provide a short list of distributors who are perceived to do a good job with products similar to your own.

Customers, of course, bring to mind end-users but for some product categories, the purchasing process will be heavily influenced by intermediaries such as consultants, retailers, advisory boards, journal editors, etc. They too must be treated as customers.

The textbook way to classify their desires is still the best: who, what, when, where, how and why do they buy?

Who?

Who purchases products similar to yours? What is their age, income bracket, educational level, mobility, etc? For retailers, will you target the supermarket chains or the independent stores? Who decides what products are sold in a given store? The store manager or a committee? Who are the distributors they favor? Who in government makes the purchasing decisions?

What?

What are the attributes Hong Kong end-users look for in products similar to yours? Prioritize them. What is the trade-off between cost and quality or cost and brand image? This is your chance to modify your pricing list, weed out items unlikely to succeed from you product line, and make any packaging modifications that might be required.

When?

What is the near-, mid- and long-term outlook on the economy by intermediary consumers, such as retailers? How might this influence purchasing? What are the seasonal sales variations for your product category? When do end-users prefer to buy? Every day, twice a week, once a month? When and how frequently do retailers expect deliveries? Delivery levels are often critical to competitiveness, inventory levels, pricing and credit terms.

Where?

Where do your target customers do their purchasing? Are they willing to go some distance to find what they want or will they take whatever is available in the neighborhood store? Do they like to buy from general stores or do they prefer to purchase products similar to your own from smaller specialized retailers? Is the store's location in an upscale part of town important? The same type of questions can be asked of retailers. Do their purchasing managers prefer to source certain products from certain countries for reasons other than price? Do they have a policy of reducing the number of suppliers to a shortlist of large distributors?

How?

What steps are required of the customer between the arousal of initial interest and the physical act of purchasing? Are the retailers highly bureaucratic or is the decision process relatively straightforward? What are the usual payment terms and return policies?

Why?

What is the most cost-effective stimulus to purchase? Editorials in specialized journals? Personalized visits? Mail order? The right shelf space in supermarket chains? Why do retailers purchase Product X out of a list of "equivalent" offerings? Better retail margins? Image? Connections?

Warning

The temptation to skip this work can sometimes seem overwhelming. Time constraints and a desire for early results may tempt you to avoid this research and trust the first distributor with whom you feel comfortable. To do so, however, reduces the distributor selection process to a game of chance, limits your room to manoeuvre during negotiation, and renders you vulnerable to sudden requests down the line from your "partner" for a reduction in price because your product is supposedly uncompetitive. Without the kind of information which can only come from personal research, you are essentially at the mercy of your distributor, which is the way he likes it. The money you initially save by making a quick and ill-informed choice will very likely be spent later as you are forced into additional concessions and travel to patch things up.

DEVELOPING A SHORT LIST

By this stage, if you have followed our advice, the process of learning more about your customers' needs and researching the competition will have also produced a list of distributors in your product category who have a reputation for doing a good job. Everything you have learned so far, every contact you have made, will be of great value when negotiations start in earnest. Now you need to have a closer look at your list of candidates.

THE "NO FRILLS" WAY

Criteria

Selection criteria are not equally important for all products. Marketing expertise, for example, is crucial for consumer products and less so for capital goods where technical know-how and after-sales servicing facilities are key. The short list of criteria used by many suppliers can be summarized as follows:

- **Financial health**: enough to ensure that bills are paid but not so much as to kill their appetite for new products.
- **Compatibility** between the distributor's existing lines and the supplier's products.
- Sufficient **sales staff** and **marketing expertise**.
- For suppliers of capital goods, adequate **after-sale servicing facilities** (technicians and equipment).
- An established **track record**.
- For suppliers interested in the PRC, proof that the distributor takes more than an opportunistic transaction-to-transaction approach to PRC trade. **Offices in the PRC** are important.
- An **enthusiastic response** to the supplier and his product.

Partnership again

This last criteria bears repeating. Notwithstanding what has been said about the limitations of "partnerships" between suppliers and distributors (see Some Fundamentals), many experienced exporters, especially those who select small distributors because they are starting from scratch in Hong Kong, will overlook certain organizational deficiencies if they sense that the distributor sales force and management are solidly behind the product. Their enthusiasm may be the result of the prestige the new product brings to the group, the avenues to other business it opens, or the good human relationships on both sides (between the owners, the supplier's sales rep and the distributor's sales force). What accounts most for small and medium suppliers breaking into the Hong Kong market is effective representation, as soon as possible. Salesmen must be out there showing the product, exhibiting it and making the calls. Until that happens, there are no sales and no cash flow.

Credit checks

The credit-worthiness of a company includes its current financial position, business reputation and experience. This information is not easily available in Hong Kong. Government bodies such as HKTDC or trade associations do supply names but will not recommend any distributor in particular or disclose credit information. This, along with the secretive nature of many small, family-owned enterprises and the lax nature of Hong Kong's disclosure laws, make much company information unavailable to the public. Three approaches are commonly used to make credit enquiries:

- **The banks** are the most common but least satisfactory avenue. The credit and trade information departments of Hong Kong's larger banks process hundreds of credit requests every day, some much more thoroughly than others. Banks are reluctant, however, to disclose which of their clients are in financial trouble for fear of diminishing their credit worthiness in the eyes of other banks, thereby lessening their own chances of recovering their loans.

- **Private credit companies,** such as Dun & Bradstreet (H.K.) Ltd. and Commercial Enquiry Services Asia, offer comprehensive company profile information for an annual fee to their subscribers.

Dun & Bradstreet, in particular, publishes two useful books containing credit information about large companies operating in Hong Kong: *Registry of Hong Kong Traders* and *Top 500 Foreign Companies in Hong Kong.* For subscribers wishing to make enquiries about smaller companies, it is best to obtain an "international introductory card" from the D&B office nearest you before appearing at the Hong Kong office.

Dun & Bradstreet (H.K.) Ltd.
12/F K. Wah Centre
191 Java Rd, North Point
Hong Kong
Tel.: (852) 516-1111
Fax: (852)562-6044

Commercial Enquiry Services Asia
A division of the CTS Group
1501 China Chem, Golden Plaza
77 Mody Road, Tsimshatsui East
Kowloon, Hong Kong
Tel.: (852) 739-9938
Fax: (852) 739- 9880

- Many exporters feel that the best, most reliable, and cheapest credit information come from other suppliers already dealing with the distributor. Such advice is best cross-checked, however, because other suppliers may not want you to distract their distributor's attention with new products.

Creative snooping

Gathering information on other aspects of the distributor's organization (i.e., the number of salesmen, its turnover rate, etc.) requires a playful tolerance of hype and obfuscation. Barging into the distributor's office unannounced or adopting an inquisitorial line of questioning in the owner's presence will get you nowhere. The best approach is to appear less informed than you really are, ask questions in a haphazard way and never write figures in front of the person to whom you are speaking. At the first opportunity, jot down the key figures: sales volumes, inventory levels, number of sales staff, number of service technicians, etc.

Some suppliers prefer to initiate contact during exhibitions when the distributor's salesmen are scattered around the room. Others think the best approach is to interview the distributor in his own premises because of the many opportunities to cross-check what has been said with staff members. All of the distributor's employees are fair game. You may find out from a receptionist, for example, that most of the staff are subcontracted to other companies, or the security guard at the door might tell you that the repair facilities do not belong to the distributor. Expect inconsistencies and do not adopt an outraged, self-righteous attitude.

THE COMPREHENSIVE WAY

In its publication, *Finding and Managing Distributors in Asia/ Pacific* (US$485), the well-known consulting firm, Business International, presents a more comprehensive method to develop a short list. Its

only drawback is that it may require more time and resources than most companies are willing to put into distributor selection. However, one option you have is to hire such firms as Business International to do the job for you.

Criteria

Business International lists 13 key criteria:

- Financial soundness and depth.
- Marketing management expertise and sophistication.
- Satisfactory contacts and relations with customers or the industry.
- Capability to provide adequate sales coverage.
- Overall good reputation and image.
- Product compatibility.
- Pertinent technical know-how.
- Adequate technical facilities and service support capability.
- Adequate staff.
- Proven performance record.
- Positive attitude toward your products.
- Mature outlook regarding future developments in market management.
- Good government relations.

Analysis

Each of these criteria is then weighted from 1 (standard value) to 5 (critical success factor). To use our previous example, a supplier exporting fast-moving consumer products might assign 5 to marketing management because it is crucial, 4 to financial soundness because inventory levels have to be kept high, and 1 to adequate technical facilities because it is less important. Someone exporting machinery would weight the criteria differently.

The candidate distributors are then evaluated according to these thirteen criteria and assigned a score for each criteria ranging from 1 (unsatisfactory) to 5 (outstanding).

The rest is simple: for a given candidate, go down the list of criteria, multiply the weighting score (1-5) with the evaluation score (1-5) and add up all his results to arrive at his total score. After doing the same for all the other candidates on your list, the three or four distributors with the highest scores can be targeted for negotiations.

MARKET RESEARCH FIRMS

Business International is just one of many market research and consulting firms with offices in Hong Kong willing to help you for a fee (see Annex 4). They range from number-crunching firms specialized in consumer purchasing behavior to more strategically- oriented firms whose strength is the comparative evaluation of the client's investment options in the whole Asia Pacific region. Not all are equally enthusiastic about accepting basic trade enquiries (the process we described in Getting the Big Picture), but will do so for between US$10,000 and US$25,000 depending on the complexity of your product and the time they are given to do the research.

Our recommendation, especially for small and medium enterprises, is to rely on these firms only for narrowly-defined questions requiring more sophisticated analysis. At the early stages, most of the statistics you require are available at very low cost from the Hong Kong Census and Statistics Department (see Annex 1). More importantly, the best market study written by someone else can never make up for the 'market sense' and network of contacts you develop by conducting your own interviews.

WEAK POINTS AND TROUBLE SPOTS

Most of the difficulties described below can be avoided or at least kept under control if the supplier has done his homework before beginning serious negoti- ations, has a preliminary promotion plan he can use to test the candidate's willingness to invest in a new product, and has protected himself with a well- drafted distribution contract.

WRONG ENTHUSIASM

In the best case scenario, the dis- tributor accepts your line because he thinks he can achieve a respectable market share for you while generating good profits for himself. However, certain distributors may be interested in your product for reasons that have nothing to do with market share.

The vampires

These are generally small distribu- tors devoid of sales planning and weighed down by a multitude of mis- matched product lines whose survival depends on securing the rights to new products as fast as older ones are taken from them. Their cash flow is generated during the highly profitable early months of a franchise when inexperienced suppliers are most will- ing to sell their goods at sampling prices or on credit. A few months later, the whole relationship ruptures when sales drop because the prod- uct's novelty has worn off and the supplier grows impatient with the dis- tributor's excuses for non-payment. Hence, the distributor's perpetual quest for new suppliers.

Less competition

A distributor, large or small, may be interested in your product because he wants to limit competition. There are two common scenarios:

- The distributor may be suffi- ciently intrigued by your prod- uct's possibilities to want to prevent other distributors from getting their hands on it. His first instinct is to sign you up as soon as possible and determine your product's real value later. To increase his flexibility, he will also want the smallest possible initial stocking order and a favorable return of merchandise clause in his contract.

- Unbeknownst to you, the distrib- utor may have the distribution rights for a competing product through a sister company. His goal therefore may be to bottle up your product in a useless dis- tribution agreement in order to protect your competitor.

Professional bidders

Some import/export firms which frequently compete for tenders are always on the lookout for "fillers" because they know that their chances of success are far greater if they can quote for all items listed in a tender. One good way to secure access to products occasionally listed in tenders and at reasonable prices is to become the authorized distributors for these products. The trick for you is to recognize the business this kind of company specializes in and grant it the limited rights of a manufacturer's representative to secure a few orders for yourself when they do win tenders. Never, however, be under the illusion that they are true distributors interested in market share for your product.

A stepping stone

Another possibility is that the distributor only wants your product because it can open the way to other franchises which interest him more. For legitimate businesses, the distributor may simply feel that your product increases his prestige or broadens his existing lines sufficiently to entice other suppliers to contact him. In other cases, your product may be used to substantiate false claims in import documents, in effect, as a cover for a far more lucrative underground business.

MARKETING SUPPORT

Distributors as a group are very wary of advertising because it is impossible to quantify its value. Consequently, promotional campaigns are often jointly planned and financed by suppliers and their distributors. With smaller distributors, other marketing problems come to the fore which may come as a surprise to some exporters new to Asia.

- Market research and planning as a discipline is alien to many small distributors used to relying on word-of-mouth information flowing through their network of friends and associates. Their tendency is to launch a product cheaply through their contacts rather than spend money on promotion. If the product does not take off quickly, they drop it.

- Small firms like to travel light, with plenty of loose change for deals and emergencies. They are often reluctant to tie up scarce financial resources to keep what the supplier thinks is sufficient inventory.

- By Western standards they sometimes seem to be nonaggressive sellers. At their worst, after an initial barrage of calls to their contacts, they can end up sitting on their stock, passively waiting for orders or even selling directly to end-users.

What emerges loud and clear from this picture is that successful exporting to Hong Kong by suppliers who choose to go through local Chinese distributors requires more careful scrutiny, some market research before selection, and close, sympathetic and ongoing marketing support once the distributor has been selected. This may cost more in terms of travel and communication expenses; however, two or three eager and well-trained salesmen pushing your product full-time can often succeed where a half dozen unmotivated salesmen from big distribution firms fail.

PARALLEL IMPORTS

Countries with high tariff barriers have to contend with smuggling; free ports like Hong Kong worry about parallel imports or parallel trade. Parallel trade refers to international commerce in goods by traders who are outside of the official channels but whose operations are entirely legal.

For example, you as a manufacturer of soft drinks may appoint an official distributor for Hong Kong authorized to import your product direct from North America. However, some enterprising Hong Kong import/export company might find out that you have licensed someone in Singapore or Mexico to produce the same drink and that he can import and sell it for a profit in Hong Kong at a lower price than your official distributor does. That is parallel importing and it is most prevalent in Hong Kong with certain cosmetics and fast-moving consumer products.

Parallel imports, if left uncontrolled, can quickly take away the distributor's incentive to work with you. The transnational distributors are financially strong enough to defend themselves in the short term. Through their superior information networks, they can learn that a container load of your soft drinks is being imported by

X and subsequently underbid him until X, who works for small margins on an L/C basis and has no storage space, is forced to sell at a loss. Smaller distributors, however, may not be able to play this game and you will lose them quickly if the problem is not promptly solved.

The only cure for parallel trade are clear and enforceable agreements with your partners worldwide; a global pricing structure which takes into account the peculiar situation of free ports; packaging which, if not unique to Hong Kong, at least avoids the styles popular in known havens of parallel trade such as the UK, Holland and Singapore; and close collaboration with your Hong Kong distributor.

IPR AND THE PRC

Hong Kong has come a long way in shaking its image as a haven for pirates of intellectual property rights (IPR). Its slow but sure integration into South China's economy may, however, lead to more IPR problems. China's long-awaited copyright law and implementation regulations which came into effect in June 1991 failed to meet the expectations of Western companies. Particularly worrisome is the weak protection it provides computer software, an issue of major concern for foreign exporters and investors.

Another leakage problem increasing in importance is the re-export of foreign products into the PRC through underground channels (see

Market Entry and the Hong Kong Middleman). This may be an issue for some suppliers who wish to enter the PRC market in a more structured way or who simply prefer that their distributors concentrate on the more demanding Hong Kong market.

PERSONNEL TURNOVER

With turnover rates between 25% and 50%, finding and keeping good employees is becoming increasingly difficult for Hong Kong firms. The problem is often more acute with ambitious salesmen and sales managers because of their frequent "representative" trips outside the office. Many of the most talented eventually hope to run their own businesses but are forced by the high cost of office space to work for someone else while moonlighting on the sly. Hong Kong is full of small entrepreneurs occupied with their own small business while sitting at the desk of a salary-paying company. Pagers, portable phones, small business centers and personal fax machines have made this game a lot easier to manage.

Large distributors have the size and structure required to maintain performance despite these losses. The same is not true with smaller firms where high turnover rates are often symptomatic of the limited possibility of advancement in family businesses (see Management Hong Kong-Style). This has obvious negative implications for suppliers who invest resources in training only to see their pupils move on to greener pastures. Training may have to be restricted to family members who have a deeper commitment to the business.

DISTRIBUTION AGREEMENTS

A CULTURE SHOCK DICTIONARY

When in Hong Kong, the aphorisms listed below should be kept in the back of your mind while, at the same time, accepting them with a grain of salt. Hong Kong is such a crossroad between east and west that many exceptions can be found. The golden rule is to keep your eyes open and adapt to the situation without losing your common sense. Hong Kong people did not get to where they are today by allowing cultural niceties to get in the way of making money.

Anger expressed publicly is a major social error. The Hong Kong Chinese consider it childish and uncivilized. Social harmony and a sense of decorum are highly valued.

Boasting is disliked. Don't belittle Hong Kong or China by making unfavorable comparisons with North America. The same goes for your own abilities and accomplishments; always remain modest, even self-deprecating.

Business cards are absolutely essential. For Hong Kong, a Chinese translation of your name on the back side of your card is unnecessary because no one will refer to you by your Chinese name. For the PRC, a translation in simplified characters is useful to have.

Business conversations with Hong Kong people can sometimes prove frustrating for confessional types anxious to "communicate". Hardly given to self-disclosure, Hong Kong businessmen often guide the conversation towards establishing relative status rather than intimacy. Do not expect the kind of instant friendships so common in North America.

Confucianism is a body of conservative social values which developed around the sayings attributed to Confucius (551-479 B.C.). For two thousand years, Confucian ideology was made the chief subject of study by China's political elite. In Hong Kong, it survives mainly as a code of family values and a basic attitude towards human relationships. Popular Confucianism means:

- leaders and elders are treated as role-models and teachers.
- loyalty to people outweighs abstract principles and law.
- a sensitivity to personal and group prestige.
- a taboo against the public expression of hostility or aggressiveness.
- a willingness to sacrifice personal wishes to social harmony.
- a fatalistic attitude towards government (the public service's responsiveness to individuals is commensurate with their social status).
- a belief in the importance of education.

- an ethic of thrift, hard work and pragmatism.
- a lack of interest in salvation and spirituality.
- a strong loyalty to family and close friends with a corresponding distrust of outsiders.
- a patriarchal family structure.

Cramped conditions are not incompatible with consumer spending. On the contrary! If a household includes four employed family members contributing money for common expenses, they can afford televisions, stereos and other modern gadgets seemingly incongruous with their housing conditions.

Criticism of someone is a loss of face for him, especially if done in front of others. Avoid it if possible. If it is absolutely necessary, do it privately, balance it with praise and be very indirect.

Dress is an expression of your status. Avoid clothes that are too informal, flashy or colorful.

Education is valued in Hong Kong and the pressure placed on children to excel academically can be gruelling.

Elderly people should be shown respect and kindness at all times.

English is the most commonly used language in Hong Kong business circles after Cantonese.

Ethics Although Hong Kong's business community is often portrayed as pragmatic and amoral, wealth creation across complex subcontracting networks imposes stringent moral values tacitly understood by all. People who miss deadlines, people whose products do not meet specifications originally promised, people whose advice turns out to be worthless are cut off. The price of staying in business is giving people what they want. The inadequate is not acceptable.

Face is the prestige and respect one has in the eyes of others, and is very important to the Hong Kong Chinese. Much of their obsession with material things and status symbols is tied in with face, not wealth for its own sake. Face is as much a group concern as an individual matter. If a group member wins a prize or commits a crime, the whole group (family, company, club, etc.) shares in his prestige or shame. Face can not only be lost or saved in Hong Kong, it can also be given. Giving face means doing something to enhance someone else's prestige among his peers. Publicly thanking someone for his help or giving a gift with snob appeal, like an expensive bottle of imported liquor, carries great weight among Chinese people.

Favors are very important. Doing business in Hong Kong is, first and foremost, a process of making friends, and favors are the currency of friendship. Money, of course, is also important but Hong Kong businessmen know that the success or failure of a venture often depends on a network of friends that money cannot buy. Favors create a sense of mutual obligation (*guanxi*) without which business partners do not feel at ease with each other (see **Networking** and **Reciprocity**).

Filial piety lies at the heart of Chinese family values. Because a Chinese father both leads and symbolizes his family, his children's identity and sense of self-worth are closely connected to his prestige and authority. Common ownership of family property is another factor preventing the emergence of the kind of psychological independence prevalent among Western children. With no independent resources and no legitimate right to oppose his father, the Chinese child emerges from twenty years of disciplined paternalism with a strong sense of debt and a respect for authority.

First names should be avoided when speaking to Chinese or British businesspeople unless you are given a clear signal that they wish to be on a first-name basis. The Chinese are especially prone to interpret excessive familiarity as a lack of respect.

Foreign businesswomen are perceived as foreigners first and women second. As long as a woman dresses conservatively and acts professionally, gender will not be a factor.

Frankness about your motivations and business goals is always appreciated by Hong Kong businessmen; they hate beating around the bush! When it comes to human relationships, however, a respect for appearances and human feelings often outweighs frankness. People prize social harmony, not always saying what they mean. Expect plenty of sugarcoating from those with whom you are not on intimate terms.

Generational differences often lead to clashes within families living in close quarters. One common flash point is child rearing. Younger Chinese parents feed their children more milk and meat (hence the taller build of many youths), and are more liberal and experimental in educational matters. Grandparents, however, still believe in traditional taboos concerning child rearing: they prefer soybean gruel to formula, and a more authoritarian approach to education.

Generosity and ostentation are a great source of prestige and face; don't be cheap.

Getting away from it all is simply not an option for most Hong Kong people. Population density and social expectations are such that even escaping to the beach or up a mountain is done in throngs.

Guanxi (the Mandarin pronunciation) literally means "connection". To have *guanxi* is to have friends who can be counted on to do you a favor and bend the rules on your behalf. Favors, of course, must eventually be reciprocated. At the psychological level, *guanxi* exists between two people when they have a sense of mutual obligation. Without it, there can be no trust.

Handing paper or business cards with both hands shows respect. Most Hong Kong businessmen, however, are sufficiently Westernized to offer their cards with one hand, and when this happens, you can, of course, do the same.

Invitations by Hong Kong businesspersons are usually to restaurants and nightclubs, not to their homes.

Money and romance are entwined in the mentality of the Hong Kong Chinese. "Separation money" may be demanded by either side when a dating couple breaks up. Money is also behind the increasingly delayed age of marriage (on average, 26 for women and 29 for men) because larger and larger sums must be put aside before a marriage can take place. Chinese weddings are very expensive even if guests are expected to bring cheques and gift coupons (wedding invitations are often dubbed "blackmail letters").

Networking in Hong Kong is not an exchange of business cards; it is an exchange of favors. A good networker is one who listens carefully to find needs he can satisfy at no great cost to himself. Can he introduce the speaker to a prospective customer? A supplier? His whole focus is on one simple question: what can I do to help this person? By working at this daily, he is developing a network of people who owe him favors.

Noise in Hong Kong is associated with crowds, home life and festivity – all positive aspects of life. From a Western perspective, Hong Kong people seem to have developed an inner barrier to cacophony. Radios are everywhere, even at beaches and mountain "retreats", and the volume is usually turned up high enough that everyone can enjoy the din, whether they wish to or not.

One-upmanship is another name for face. Hong Kong people work hard in the belief that if the next guy can do well enough to own a Mercedes, so can they. Keeping up with the Wongs is not enough; you must overtake them!

Patience is highly valued. To flare up when things go wrong is to display a lack of self-control which the Chinese find childish.

Paying for meals is done by the person who did the inviting. "Going Dutch" is never done. If no one in particular made the invitation, the bill is fought over. Paying is considered a matter of pride.

Price shopping in Hong Kong is so obsessive it has been compared to sex in the West. The price of goods is a subject of eternal fascination and everyone knows where an item can be purchased cheaper. In most fields, the discounted item wins the day, a fact of which the most powerful retail chains are only too aware. Forty-five percent of everything Park 'N Shop sells is sold on promotion, one of the highest percentages for supermarkets around the world.

Psychological withdrawal, the ability to switch off into a personal bubble, is a survival skill many Hong Kong people have developed to remain sane in what can be a highly demanding and competitive society. The average tenement dweller ignores his neighbors when he leaves in the morning, barges into the train to commute to the office, and sees nothing wrong in shouting into a cellular phone in places usually regarded as peaceful and quiet. He doesn't care what the people next to him are thinking and assumes they are equally capable of distancing themselves from the loud music, smoke, strident voices and beeping pagers around them.

Queue jumping is a national sport in Hong Kong, part of its "mind your own business" ethos some foreign visitors find shocking. Expect people to rush into elevators and trains without giving a thought to those who wish to get out.

Reciprocity governs the exchange of favors. Do not give gifts so expensive or favors so great that the recipient is unable to reciprocate. To do this is to invite refusal.

Recommendations are not given lightly in Hong Kong for fear of cheapening the currency. If someone's introductions give you access to useful contacts, you now owe that person and are obliged to do everything possible to justify his good opinion of you in the eyes of his contacts.

A **refugee mentality** underlies the ruthless edge of Hong Kong life. Its insecurity, its extraordinary faith in wealth as the royal path to personal security, and its matter-of-fact indifference to the host society reflect the world view of an immigrant population, many of whom fled harsher conditions in China.

Respect corporate chains of command. Once you have a contact in an organization through whom you have made proposals or arranged meetings, do not try to circumvent him unless you are explicitly told to do so by someone at a very high level. No one likes an outsider who mucks up organizational hierarchies.

Revenge is a major theme in Chinese literature. Embarrassing a Chinese person in public, even inadvertently, will lead to retribution of one type or another. The Chinese do not usually show anger; rather, they get even. Face is very important to them.

Social harmony should be preserved at all costs, and may require small lies and indirect ways of communicating your message.

Status is very important. Acknowledging other people's status simplifies social relationships as does acting according to one's own status.

Taboos and omens are taken seriously by many Hong Kong Chinese. Anything associated with death is avoided as unlucky. Because white, not black, is the color of death, white flowers or gifts wrapped in white paper are only given at funerals. Watches or clocks are also considered unlucky business gifts because timepieces denote a "count-down" towards death. Birthdays, the lunar New Year and wedding days have a host of rules to ensure good luck. On the whole, young people and the well-educated care less about these taboos than the elderly.

Titles can be misleading because many Hong Kong companies endow certain employees with higher-sounding titles than their counterparts in Europe or the US. This is frequently the case in distribution and retailing where a sales person might be called a marketing executive to give him a certain amount of "face" when dealing with high-level customers.

Touching in public The basic rule is not to touch members of the opposite sex in public, except for handshakes. Common sense is the best guide; Hong Kong is far more Westernized than both Taiwan and the PRC in this matter.

Wealth To flaunt one's wealth is far more acceptable in Hong Kong than in Taiwan or the PRC. But spending money is not synonymous with making it and many Hong Kong people are caught between the desire to parade their affluence and the need to spare their bank accounts. This explains, in part, the market for counterfeit prestige watches, dummy cellular phones and the urge, in certain extreme cases, to gamble for high stakes.

Young executives, however technically capable, must be particularly careful when dealing with older Chinese businesspeople in Hong Kong. Avoid anything that could be interpreted as a sign of arrogance, such as boasting, speaking to someone new on a first-name basis, belittling the competition or the competence of Hong Kong people, etc. There is a cultural assumption in Asia that age connotes experience and that youth has everything to learn.

NETWORKING AND NEGOTIATIONS

Prospecting a short list of good distributors before serious negotiations begin (see Developing a Short List) requires certain basic networking skills we describe in this chapter. As for the negotiation process itself, we will concentrate on those aspects of the distribution agreement which are most important from a business standpoint. The two chapters which follow, "Elements of a Distribution Agreement" and "Taxation in Hong Kong", examine these same agreements from a legal and tax perspective, respectively.

EXPLORATORY INTERVIEWS

Secretaries and receptionists

The best way to be taken seriously is to be referred by your country's trade commission in Hong Kong because distributors wish to remain in good standing with organizations which send them a steady stream of potential suppliers. Even without an introduction, however, Hong Kong is such an open city that cold calls can potentially get you very far.

Unless you know the person to whom you wish to speak, expect to waste some time being shunted back and forth between different departments. Receptionists, often poorly paid young women anxious to move on to something better, usually know little about the company they serve beyond the list of names and extension numbers placed in front of them. Requests to speak to the manager may therefore be greeted with confusion. Language, too, can sometimes cause difficulties: about 50% of secretaries have problems understanding English.

The hook

Once you reach the person you want, make sure to get your hook in early! If your company is unknown in Hong Kong, mentioning its name is of no use. Worse still are wild claims about your influential connections or how you are the best at something. The more useful approach is to state that you represent a Canadian or American company, that you have a toothpaste, pacemaker, etc., which sells very well in North America and that you would like to meet him to get his advice about the Hong Kong market. Be sure to state plainly whether you are exploring the market or actively into distributor selection. Frankness in these matters will certainly not hurt you in their eyes; Hong Kong businesspeople hate to beat around the bush!

Some background

The character of exploratory interviews can vary a great deal depending on the size of the organization you contact. Large transnational distributors, for example, combine tough financial controls with decentralized divisional profit centers where tasks are highly differentiated. Within a single division of JDH Trading Limited (Inchcape Pacific), for instance, the same product will be managed by a sales manager who oversees a sales force and a merchandising team, and a marketing manager responsible for all promotion and advertising activities. Both may have very different perspectives on your product's market potential. Another characteristic of large organizations is that their managers are comfortable with Western marketing concepts and less worried about signing you on (or dismissing you) immediately. As a consequence, communication is often easier, and certain distributors interested in your product may even allow you to accompany a sales team for a day.

Small and medium family enterprises present a very different picture. They can be divided into three groups, each a little more open than the last. First are the archetypal Chinese family enterprises we described in "Management Hong Kong-Style". In its purest form, this traditional management style is far more prevalent in manufacturing than in distribution where it survives among some wholesalers. Second are enterprises run on a day-to-day basis by Western-trained sons but where a father-founder figure retains veto power. Third are companies recently founded by young entrepreneurs who understand modern management techniques. Open or not, managers of small enterprises who agree to meet you, live in a world of rush orders and last-minute changes where they must take charge of every detail. Compared to the managers in large organizations, they will probably be more secretive, have less time, and be less capable of giving you a sophisticated overview of the market. On the other hand, they are often better placed than other managers to provide insights into some of the ground-level distribution problems they face because they deal with them personally every day.

Arrival

While commuting from one meeting to another is relatively easy in Hong Kong, avoid stacking meetings too close to each other because no one is impressed if you must exit quickly in order to meet someone else. It gives the impression that you are another one of those "two days here, two days there" fly-by-nights.

Being ten to fifteen minutes late is considered acceptable because of traffic problems in Hong Kong. However, not showing up at all without some forewarning is very bad form, especially if the meeting was organized by a mutual acquaintance. Do everything possible to avoid this situation and apologize promptly or you will be burning bridges.

Appearances

Your general appearance is both an expression of your status and your respect for the host's status, all very important in Hong Kong society. Being well-dressed and groomed (not necessarily expensively) is a must. The quality of your supportive material must also be up to standard. Finally, while your hotel need not be the Peninsula, it should be a high caliber one, at least during the early contact and negotiation stage. Later on, when they have come to know you, the hotel you stay in will matter less.

Meetings

Upon arrival, you will either be ushered directly into your host's office or, more likely, into a boardroom where you will be offered something to drink and asked to wait. If this waiting period stretches to fifteen minutes, no slight is intended. Hong Kong managers often must respond to problems on the spot or answer long-distance calls.

Giving your card is the accepted way to start a meeting. Presenting it with both hands shows respect and is the safe thing to do unless you notice your host handing his card out in the Western way. If an older person is present in the room, be sure to show him some respect since age is still important in Hong Kong.

Since you have only 25-30 minutes at the most, your host will be very grateful if you get to the point of your visit as quickly as possible. Explain to him who you are, where your company is based, its size and importance, the bank it uses (this gives it an additional aura of stability), your basic product, why Hong Kong is important to you now, what you hope to gain from this meeting, etc. Having these details written on a sheet of paper and handed out or faxed before a meeting can be very useful. When meeting, get to your questions quickly. Be sure to stress areas of possible advantage to your listener and try your best to get him to agree to a second meeting or at least a phone call at some future unspecified time if you should need additional information.

NEGOTIATIONS

Shadows

Never give a distributor the impression he is your only choice. Indeed, if you have done your groundwork properly, you should have three or four good candidates; even if one stands out, try to create the appearance that you are seriously considering other options.

Never allow a single distributor to meet you at the airport or otherwise control your schedule. Allude occasionally to competing distributors in you discussions, carry their business cards and literature in your briefcase, mention other meetings, real or imaginary. Look like a good prospect who should not be taken for granted.

Preparation

If the distributors on your short list are good and your product is new to Hong Kong, you are in a buyer's market and must be prepared for plenty of skepticism. Good distributors fear the loss of focus which comes from unprofitable diversification and the first questions you can expect to be asked will center around profit margins, potential sales volumes, product compatibility with existing lines, inventory levels and after-sales service requirements.

Consequently, you must have a well-rehearsed sales pitch, one that gives a full and convincing picture without allowing your audience to get bogged down in misunderstandings. Some suppliers even suggest that inexperienced exporters deal with their best candidate last because the first presentations are bound to reveal many unanticipated shortcomings in their assumptions. Do this and, by the last one or two candidates, you should have a good answer for almost any objection.

Your *spiel*

After ferreting out who your competitors are, what channel mechanisms are most appropriate for your product, and who your target customers are, this is where the pre-negotiation groundwork really pays off.

With respect to the **market**, list your target customer groups, prioritize them and talk about the most cost-effective way they can be reached during the first year. Give preliminary sales figures.

This leads you into the realm of **profitability.** To minimize confusion, have prepared examples on hand which factor in the discounts, tender prices, slotting fees, kickbacks and commissions which make Hong Kong go round. Be realistic and define your terms in order to avoid getting bogged down in disputes where one

side speaks in terms of markups and retail prices and the other about gross margins and net prices.

Inventory levels

Inventory levels are best discussed during the final stage of the negotiation process, after you have both agreed on a sales promotion plan for the first year (see Launching Your Product). Indeed, it is best to keep details about training, payments, returns, activity quotas, etc., until later, when management has had an opportunity to assess the product's potential in the light of your joint sales promotion plan.

Negotiations

Negotiating in Hong Kong involves few cultural surprises for North American exporters. English is used both in discussions and in the written contract. Negotiation teams are small and proceed with a minimum of ritual. Frankness is expected, even for sensitive issues, such as negative credit checks or 1997. The pace, however, can be very fast for some and you should not hesitate to slow things down if you are not comfortable with the way discussions are going.

Expect challenges to all of your assumptions, especially your prices. According to Donald and Rebecca Angeles Hendon, the authors of *World-Class Negotiating: Dealmaking in the Global Marketplace*, a survey questionnaire revealed that the nine favorite tactics in descending order of frequency used by Hong

Kong executives when asking a seller to lower his price are as follows:

- Confront the seller with the competitor's prices (net or list).
- Start with high initial demands in order to lower the seller's expectations and create room for future concessions.
- Probe his knowledge of the market and discredit his assumptions.
- Confuse the issues by concentrating on percentage figures instead of absolute figures, costs per unit not total costs, and rounded figures instead of exact figures.
- Set up a credible deadline which is not in the seller's advantage.
- Present the seller with a single, take-it-or-leave-it option.
- If the seller's position is strong, emphasize to him how much he has to lose (services, inventory levels, whatever) if he demands too much.
- Nibble away at the seller with a thousand small concessions before and after the contract is signed, concessions which individually cost little but collectively are interesting.
- Tell the seller you are financially constrained by some credible larger force.

The effectiveness of these tactics in the final analysis will depend on two things: your knowledge of Hong Kong's market as it affects your product and the number of alternative distributors on your short list who have shown interest in your product. Both

only come to those who do the groundwork before attempting to negotiate anything; otherwise, a seller is at their mercy.

Lawyers

Lawyers are best kept out of the negotiation process until the end when draft agreements have materialized. Their presence at the beginning is, at best, a waste of their time because the situation is too fluid on which to base a legal opinion. At worst, their presence communicates your distrust to the other side. Hong Kong managers of small and medium enterprises, in particular, hate legalistic jargon even if they agree that contracts are a necessity. Keep the lawyers for the end.

The phone

Phones are best used to arrange meetings. Never negotiate on the phone, only in face-to-face meetings.

Negotiating in the PRC

Although negotiation with end-users is the distributor's responsibility, the presence of a capital equipment supplier is often necessary in PRC negotiations because of their special after-sales service and training requirements (see Market Entry and the Hong Kong Middleman). In such cases, the basic rule you must follow is to leave all lobbying efforts to your distributor: his expertise in these matters is the reason you chose to go through him. Do not make any promises which have not been agreed to earlier on with the distributor.

Worse still, do not make a mess of everything by thinking "China... connections... corruption" and handing out bribes to everyone. Bribes only attract more bribes and make your distributor's work more difficult.

The best source of information on the protocol side of negotiating in the PRC is Scott D. Seligman's book, *Dealing with the Chinese*.

INVITATIONS AND MEALS

Gifts

Gift-giving is not part of Hong Kong's business culture. Things are, of course, different in the PRC and Taiwan. In Hong Kong, the only gifts that matter are good business opportunities.

Meals

Unless time is pressing, meals are not usually a venue for serious negotiations. Breakfasts are a private affair, not usually a social occasion. Lunches can be arranged anytime with a minimum of forewarning. Dinners, however, are more memorable and meaningful. Social life is conducted in hotels and restaurants; only very intimate friends are invited to homes.

The main point of a meal together from the distributor's point of view is to know you informally, probe the extent of your network and understand your true intentions. Drinks in girlie bars push the process even further.

Unlike the PRC, there are no special seating arrangements; the only rule is straightforwardness and informality. Conversation is very business-oriented and your participation will certainly be made easier if you make a point of reading the business section of the *South China Morning Post* while you are in Hong Kong. One subject which is sure to bore everyone because every visiting foreigner brings it up is what will happen to Hong Kong in 1997. The fact is, no one really knows; all they have is hope that Beijing will ultimately respond to reason and self-interest. Whatever contingency plans individual managers may have arranged (a topic you should never bring up), the view of most Hong Kong businesspeople is that optimism about the future is a prerequisite for establishing long-term business relations in Hong Kong. Try not to sound too pessimistic if the subject comes up informally. Negotiations, of course, are different. If 1997 directly affects the deal you are considering, do not hesitate to bring it up and discuss your worries frankly.

Two things often annoy Western businesspeople invited to such meals. The first is the use of cellular phones during meals. Hong Kong professionals usually carry them along and accept the fact that meal conversations may have to be interrupted by in-coming calls. The second is the frequent switch to Cantonese when you are sitting at a table with more than one Hong Kong person. The fact that you don't understand what is said is not considered impolite.

"Going Dutch" is never done. A meal is paid by the person who made the invitation. If no one in particular made the invitation, it is good form to insist on paying or at least to extract a promise that you will pay the next time around.

THE CONTRACT

Over-rated

Perhaps because contract drafting absorbs a good deal of time during the negotiation process, exporters again and again make the mistake of ascribing intrinsic value to distribution agreements. The core of any productive cooperation between you and your distributor is his willingness to work hard selling your product, and what motivates a distributor is your product's profitability. No contract, however iron-clad, can by itself rekindle that motivation if it is lost. What it can do, however, is make a separation easier.

The simple truth therefore is that distribution contracts are divorce agreements, not marriage contracts, and function best if they give the supplier an easy exit when things go wrong and protect him during the termination process.

Two rules

First, **be specific.** Define in quantitative terms what you mean by performance standards, activities, payments and credit terms. List the names of the people who will be selling your product. Quantify what you mean by reasonable access to staff for training purposes, etc. The more specific the contract, the easier it is to terminate if the distributor does not perform.

Second, **reserve a right of veto** in cases where distribution work is subcontracted to others **and an opting-out clause** in case your distributor is taken over by another company you do not want to work with.

Exclusivity

Hong Kong is such a small, concentrated market that exclusivity is usually granted without question during the first years. Exclusivity, however, can be a very thorny problem with respect to the PRC. The advice of most experienced PRC traders is to do what the Japanese do: avoid giving exclusivity for the whole of China to any single company. If the distributor insists on it, ask him specifically what promotion activities he intends to organize in each of China's provinces. Almost certainly, he will focus his efforts in one or two cities or provinces and ignore the rest. Why then should he insist on exclusivity over parts of China that do not interest him?

ELEMENTS OF A DISTRIBUTION AGREEMENT

STEVENSON, WONG & CO.

This is an overview of distribution agreements in Hong Kong from a legal perspective. The author of this section is the Hong Kong-based law firm Stevenson, Wong & Co., the Hong Kong member of INTERLAW, an international association of law firms (see further Annex 5).

INTRODUCTION

Those involved in the sale and distribution of goods are often described as, or are viewed as, acting in an agency capacity on behalf of the supplier of the goods concerned. However, despite the fact that a distributor may have limited agency functions or carry out services that are typically carried out by an agent, the legal position of a distributor is, in reality, quite different from that of someone who enjoys the position of an agent.

The most fundamental difference between an agent and a distributor is that whereas an agent may enter into negotiations and contracts of sale on behalf of a supplier, its principal, a distributor purchases goods from the supplier outright, and then resells them to its customers under a separate contract of sale. In a distributorship arrangement, there are two distinct and separate contracts and there is no direct contractual relationship between the supplier and the customer who purchases goods from the distributor.

The fact that a distributor deals with both the supplier and the customer as principal, rather than acting as an intermediary between them, has a major effect on its legal position with respect to those parties. Whereas an agent can bind its principal and avoid personal liability, a distributor will be directly liable to the customer for any breach of the contract of sale entered into between them. Accordingly, in Hong Kong a distributor may be liable to the customer for the breach of any of the terms implied into that contract pursuant to the Sale of Goods Ordinance (Cap. 26) concerning, for example, merchantable quality or fitness for purpose, despite the fact that it has no involvement in the manufacturing process and has not been negligent in any way. On the other hand, if the supplier is the manufacturer of the goods concerned, it will only be liable to the customer if it is negligent in manufacturing the goods and if that negligence causes damage to the customer, or under the terms of any manufacturer's guarantee that may be given and in operation.

Although a distributor may have greater control over its time and efforts and a higher degree of discretion in such areas as price determination and advertising, it is also more likely to bear the risk of stock loss and slow customer demand. It is also probable that a distributor will be solely responsible for the cost of running its business, including any staff expenses that may be incurred.

Unlike an agency, a distributorship arrangement derives solely from contract law and turns almost exclusively on the terms of the particular contract that is entered into. In addition, although the appointment of a distributor is a relatively common commercial occurrence, the circumstances which may give rise to such an arrangement and, as a result, the precise nature of the relationship formulated, may vary widely. For these reasons, distribution agreements require careful and precise drafting and it is essential that those undertaking the preparation of such contracts clearly establish the particular requirements and intentions of the parties involved. It is hoped that this chapter will provide some useful assistance to those involved in the negotiation and preparation of distribution agreements concerning Hong Kong.

GENERAL ELEMENTS OF A DISTRIBUTION AGREEMENT

Although the terms of a distribution agreement may vary widely from one case to another, a relatively comprehensive agreement will normally contain provisions relating to the following:

- Description of the products to be distributed
- Description of the distributor's territory
- Nature of the relationship between the parties
  -exclusivity of supply
  -exclusivity of purchase
  -assignability/appointment of sub-distributors
- Obligations of the supplier
  -supplier support and training
  -delivery of products
  -acceptance of orders
- Obligations of the distributor
  -good faith/best endeavors
  -compliance with instructions of the supplier
  -minimum order/purchase quantity
  -keeping and storage of stock
  -sales reports and other information
  -advertising and promotion of the products
  -terms and conditions of sale
  -installation and after-sale service
- Use and protection of the supplier's intellectual property
- Price of the products and payment of the supplier
  -determination of price
  -deposit
  -discounts and penalties

- Title to the products supplied
  - -reservation of title
  - -risk
- Duration and termination of the agreement
  - -termination events
  - -loss of exclusivity
  - -disposal of stock
  - -payment of invoices
  - -compensation
  - -restraint of trade
  - -confidentiality
- Warranties and undertakings
  - -quality and fitness of the products
  - -manufacturer's warranty
  - -trademark registration
  - -indemnities

THE ELEMENTS DISCUSSED

Products

Clearly defining the product or range of products to be distributed is essential. Obviously, the supplier may not wish all of the products with which it is involved to be distributed in the designated territory by the same distributor, or indeed any distributor at all. Further, by clearly defining which products are to be the subject of the agreement, a supplier may be able to engage more than one "exclusive" distributor in the territory, therefore taking full advantage of each distributor's particular expertise and knowledge. From the distributor's point of view, in light of any undertaking that he may be required to give to the supplier with respect to its right to purchase the products from any other source, a clear description

of the relevant products will assist the distributor in determining exactly which other products it is permitted to trade in. Provision should also be made in the agreement for any new products with which the supplier may deal in the future and for any product alteration and supply discontinuance that may occur.

Territory

Another important aspect that must be clearly determined in a distribution agreement is the territory in which the products are to be distributed. If the proposed territory includes a number of countries, the applicable laws of each of those countries should be carefully considered. Although not an issue in Hong Kong, any provision in the agreement relating to territorial, supply and/or purchase exclusivity may be seen as having an anti-competitive effect and may result in the proposed distribution relationship being deemed to offend any competition or trade practice laws that may be in operation in the territory, or in any section of it. Other laws that should be addressed, particularly by the distributor, include those relating to tax (both sales and income), import duty and control, consumer protection and intellectual property. The distributor should also determine whether it is required to obtain any governmental or regulatory approval in order to operate in any part of the territory.

As with the definition of the products to be distributed, provisions should be made in the agreement for any possible territorial changes that may occur.

Nature of appointment

Although a distribution agreement may place many restrictions and obligations on the distributor, the most important provisions are those relating to the appointment of the distributor and the nature of the rights granted to it.

The supplier may appoint the distributor on either a "sole" or "exclusive" basis. Under a sole distribution arrangement, it is generally understood that the supplier will be prohibited from supplying any other distributors with the products in the territory but will still be able to sell the products there itself. Conversely, an exclusive appointment will, in most circumstances, be taken to mean that the supplier is prohibited both from supplying other distributors and from itself selling the products in the territory. However, due to the fact that the precise meanings of these terms have become confused through improper and indiscriminate use, it is advisable to clearly spell out the exact nature of the distributor's appointment and any limitations on the rights of the supplier that are to apply.

Another element of exclusivity that may be provided for in a distribution agreement relates to the purchase of the products from the supplier. The distributor may undertake to be supplied with the products only by the supplier. Such an exclusive purchasing requirement is often referred to as a "solus" term.

As was mentioned earlier, any restriction on the right of the supplier to use other distributors or on the right of the distributor to purchase the products elsewhere will have implications for any competition or trade practice laws in operation in the territory. As a result, when contemplating the exact nature of any proposed appointment, close regard should always be given to such laws.

Finally, a distribution agreement should also refer to such things as the assignment rights of the parties, the ability of the distributor to appoint sub-distributors and the title to be used by the distributor in carrying out its business.

Obligations of supplier

The supplier's most important obligation in most arrangements of this nature relates to the exclusivity of the distributor's appointment. However, the supplier may also undertake to do, or refrain from doing, a number of things in order to assist the distributor in carrying out its various obligations or to protect more fully the rights granted to the distributor under the agreement.

If the products to be distributed must be installed, are complex, or require specific instruction for their use, the supplier may agree to provide staff training and technical assistance to the distributor, to hold seminars and by the distributor's customers, or even to arrange for a suitably trained member of its own staff to assist the distributor in its business for a certain period of time. The exact nature of any assistance to be provided to the distributor and the party who is ultimately to bear the cost of that assistance should be clearly stated.

The supplier may also agree to assist the distributor with its marketing of the products, particularly if the supplier is already involved in widespread marketing activities outside the territory or wishes the products to be promoted in a particular way. As a result, the supplier may agree to provide the distributor with brochures, posters and other promotional material and information.

Given that the distributor purchases the products from the supplier outright and then resells those products to its customers under a separate and enforceable contract of sale, the obligation of the supplier to deliver the products required to a designated place by a certain time is also of great importance and should be clearly described. In order to guard against the possibility of not being able to deliver the products within the time period contemplated, the supplier may stipulate that it is under no obligation to accept any order placed with it by the distributor and that until it does accept any order it will not be bound in any way.

Obligations of distributor

It should be remembered that, unlike an agent, a distributor owes no fiduciary duties to the supplier. Therefore, the distributor may expressly undertake to the supplier to act in good faith and to use its best endeavors to carry out its operations under the agreement properly and (depending on the exclusivity of its appointment) not to deal in competing products or with the supplier's competitors.

As was mentioned previously, the distributor may be subject to the control of the supplier to the extent that the relationship between them closely resembles one of principal and agent. The distributor may be bound to comply with the orders and instructions of the supplier with regard to the marketing and distribution of the products, or to provide the supplier with regular sales reports, up-to-date customer lists or copies of its accounts. The distributor may also be required to make demand forecasts, to maintain certain minimum order/purchase levels or to keep a prescribed minimum level of stock and spare parts.

If the distributor is to be responsible for the marketing of the products, the supplier may require that all promotional materials and ideas be approved by the supplier before being used, particularly if that use directly affects any of the supplier's intellectual property rights or if the distributor has agreed that the products will be marketed in a particular way. The obligation of the distributor that may arise with respect to the supplier's intellectual property rights and to payment are discussed below.

Finally, the nature of the product or products concerned and the corresponding support of the supplier will normally dictate whether or not the distributor agrees to carry out installation and after-sales service.

Intellectual property

If the supplier owns any intellectual property rights, such as trademarks, in the territory with respect to the products, the agreement should clearly provide for the authorized use and protection of those rights by the distributor. Although the products themselves are purchased by the distributor from the supplier, the ownership of any intellectual property rights subsisting in the distributor's territory with respect to the products remains with the supplier. The supplier will want the distributor to use these rights in such a way as to maximize the success of the distributor's agreed operations. However, the supplier will also wish to maintain and protect its property by controlling the use of these rights by the distributor.

In the agreement, the distributor may be required to expressly acknowledge that, beyond the terms of the agreement, it has no rights in respect of any trade names or trade marks used by the supplier in relation to the products. The distributor will normally undertake not to alter in any way the supplier's trade marks, the products or the packaging of the products. In addition, it may agree not to use any of the supplier's intellectual property except in connection with the carrying out of its obligations under the agreement, or without first obtaining the consent or approval of the supplier to that use.

A distribution agreement will normally contain a number of provisions aimed at controlling the use of the supplier's intellectual property by the distributor. However, given that the distributor operates in the territory and therefore may have a greater awareness of the activities of other suppliers and distributors in the territory, the supplier may want the distributor to assist in protecting its intellectual property rights from infringement by third parties. The distributor may agree to notify the supplier of any actual or potential infringement that comes to its notice and to take such steps as the supplier may specify in order to protect its intellectual property rights.

In Hong Kong, if a foreign supplier has any registered trade marks which are to be used on, or in relation to, the products and if the Hong Kong distributor is to be involved in the packaging or finishing of the products and, therefore, the application of the trade marks, provision should be made in the agreement for the distributor to enter into a registered user agreement.

Price and payment

The price to be paid by the distributor for the products to be purchased from the supplier may be determined in many ways, depending on such factors as the term of the agreement and the expected cost of supply. In shorter term arrangements or where the costs of supply are expected to remain relatively stable, the price may be expressly set out in the agreement. In longer term arrangements or where supply costs are likely to be more volatile, the price may be determined by reference to price lists to be issued by the supplier from time to time or by using some prescribed formula. If prices are able to be varied, the distributor is likely to require reasonable notice of any change, to give it time to make suitable adjustments to its own operations.

It should be clearly stated whether the supplier's prices are inclusive of any value added or other sales tax that may apply, and whether any transport costs are covered.

The supplier may require that a deposit be paid by the distributor when orders are placed or that all payments be made in a particular currency. As to the timing of the payments, the distributor may receive a discount for promptness or incur a penalty, or even entitle the supplier to terminate the agreement, if it is guilty of late payment. As an incentive to the distributor, a discount or rebate may be given for bulk purchases or if the distributor purchases more than a certain quantity of the products in a certain period.

Title

It was mentioned earlier that the main difference between an agency relationship and distribution relationship relates to the title of the products supplied. Where there is an agency relationship, title to products never passes to the agent. However, a distributor will hold the position of a purchaser under a contract for the sale of goods.

Given that the supplier will, therefore, face the same risks as any other seller of goods, in order to protect itself more fully against loss resulting from the insolvency of the distributor, the supplier may insist that retention of title provisions be included in the distribution agreement. Typically, these provisions will stipulate that title to the products supplied will not pass to the distributor until the supplier receives payment in full for those products and until that time the distributor will be deemed to hold the products as bailee for the supplier. The supplier may also require the distributor to store the products separately and to mark them clearly as the supplier's products, or to open a separate bank account into which the proceeds of the sale of the products must be deposited.

The supplier and the distributor will normally agree that risk is to pass upon delivery (or collection) being effected and the distributor assuming physical control over the products. If this is the case and if title is retained by the supplier, the distributor will still be liable to pay the supplier in full for any of the products lost or damaged while in its possession.

Duration and termination

A distribution agreement may remain in force for a set period or until terminated by one party giving notice to the other in a prescribed manner.

However, the parties (particularly the supplier) will often wish to provide for the termination of the agreement at short notice upon the happening of certain events. From the supplier's point of view, given that it is primarily interested in maximizing the commercial rewards to be achieved by appointing a distributor in the territory, it may insist that it be given the right to terminate the agreement should the distributor fail to reach certain pre-determined target order/purchase levels, or should the distributor experience financial difficulties. The agreement may also provide for termination by the supplier in the event that there is a change in the management or control of the distributor, or that the distributor challenges the validity of any of the supplier's intellectual property rights.

Depending on the nature of the distributor's appointment, in addition to the supplier being given the option to terminate, the supplier may also be able to withdraw the distributor's right of exclusivity should the distributor fail to maintain a certain level of performance.

In the event that the agreement is terminated for whatever reason, as the distributor would no longer be entitled to promote or distribute the products in the territory, provision must be made for the repurchase or the disposal of any stock and promotional material in the hands of the distributor and for the cancellation of any registered user agreements entered into with respect to the supplier's intellectual property. Obviously, if title to products for which payment has not been made is expressly retained by the supplier, those products may be repossessed. Further, it is normally agreed that upon termination, all outstanding unpaid invoices

rendered by the supplier with respect to the products will become immediately payable and that, in addition, immediate payment will be required for products ordered prior to termination but for which no invoice has at that date been rendered.

In the event that the agreement is terminated for reasons other than the distributor's poor business performance or failure to adhere to the terms of the agreement, the supplier may wish the distributor to receive some form of compensation. However, many distribution agreements expressly state that the distributor, in the event of termination, will have no claim whatsoever against the supplier for compensation for loss for distribution rights.

Finally, in order to protect its operations in the territory after the termination of the agreement, the distributor may be made subject to non-competition or restraint of trade provisions. In addition, the obligation of the distributor to keep certain information, such as customer and price lists, confidential, both prior to and following the termination of the agreement, is also normally included.

Warranties and undertakings

Depending on the law under which the agreement will operate, the distributor may require the supplier to give some undertaking as to the quality and fitness of the products to be supplied. The distributor may also require the supplier, if the supplier is the manufacturer of the products, to give a manufacturer's guarantee to the distributor's customers. The supplier may also be required to warrant that any trade marks that are described as being registered in its name are in fact properly registered.

The distributor is likely to require the supplier to offer it an indemnity against all losses suffered and expenses incurred by it as a result of the negligence of the supplier (or its agents, etc.), or an infringement of the intellectual property rights of any third party. From the supplier's standpoint, it may ask to be indemnified by the distributor against any losses suffered and expenses incurred by it as a direct result of the misconduct or default of the distributor.

CHECKLIST

As was mentioned previously, the relationship between a given supplier and a given distributor depends almost entirely on the terms of the distribution agreement that is entered into between them. The particular requirements of the parties may vary greatly depending on such factors as the nature of the products to be distributed, the size of the relevant market to be supplied, the supplier's knowledge of that market and the resources and expertise of both the supplier and the distributor. In addition, the applicable laws in both the manufacturer's jurisdiction and in the territory to be supplied, particularly in the areas of tax, consumer liability and competition, may also have a bearing on the nature of the proposed relationship between the parties. It is therefore vital when preparing to draft a distribution agreement to obtain a clear picture of the actual requirements of the relevant parties. The following checklist is intended to be of assistance in this regard.

Products

- Can the products be clearly defined?
- Do the products require installation, training of sales staff, operation manuals, after-sales service?
- Are the products already available to customers in the proposed territory?
- Does the supplier currently advertise and promote the products in a particular way?
- Will the range of products change?
- Is the supplier to be able to alter the products?

Territory

- Can the territory be clearly defined?
- Does the territory include more than one country?
- Are there any applicable competition laws in any part of the territory?
- Must the distributor obtain any governmental or regulatory approval in order to operate in any part of the territory?
- Is the distributor to be responsible for obtaining any import/ trade licences that may be require?

- Must the distributor maintain particular sales outlets within the territory?
- Will the territory be extended?

Nature of appointment

- Is the distributor only to purchase the products from the supplier?
- Is the distributor permitted to trade in non-competing products?
- Does the supplier intend to use other distributors for the products in the territory?
- Is the distributor to be responsible for marketing the products in the territory?
- Will the distributor be able to take orders from outside the territory?
- Can the supplier sell directly to customers in the territory?
- Can the distributor appoint sub-distributors?
- Are the parties to be free to assign their contractual rights and obligations?
- How is the distributor to be described?

Obligations of supplier

- Will the supplier provide staff training or other assistance to the distributor?
- How and when are the products to be delivered?
- Will the supplier pay transport costs?

- Is the supplier obliged to accept all orders received from the distributor?
- Is the supplier to be bound prior to accepting orders?

Obligations of distributor

- Must the distributor comply with the instructions and directions of the supplier?
- Must the distributor report to the supplier with regard to sales in the territory?
- Will the distributor be required to forecast demand?
- Are there any minimum order/purchase requirements?
- Will the distributor be required to hold a certain level of stock?
- Will the distributor be permitted to offer credit to customers?
- Is the distributor to pay all its business expenses?
- Are certain terms and conditions to apply to sales made by the distributor?
- Is certain information to remain confidential?
- Will the distributor be solely responsible for fixing the prices of products sold to customers in the territory?

Intellectual property

- Does the supplier have any intellectual property rights with respect to the products in the territory?

- Is there to be any restriction on the distributor's use of these rights?
- Is a registered user agreement required?

Price and payment

- How are the prices of the products supplied to the distributor to be determined?
- Are price lists to be supplied?
- Is a formula for price setting to be used?
- Is there to be any discount for early payment or for large purchases?
- Is a deposit required?
- Will the distributor be allowed credit?
- Are the prices inclusive of any sales or other taxes?
- Are payments to be made in any particular currency?

Title

- When is title to pass to the distributor?
- Are servation of title provisions required?
- Is a separate bank account required?

- Will the distributor be required to store the products in a particular way?

Duration and termination

- Is the agreement to be for a fixed term?
- Is provision to be made for renewal?
- If determinable by notice, what is to be the notice period?
- Is the agreement able to be terminated upon the occurrence of certain events?
- Is termination to be performance related?
- Will the distributor receive any form of compensation upon termination?
- Is the distributor to be subject to restraint of trade provisions?
- Will the supplier repurchase unsold stock?

Warranties and undertaking

- Are any warranties or indemnities to be given?
- Will the sale of Goods Ordinance (Cap. 26) apply?
- Will the supplier issue a manufacturer's guarantee?

TAXATION IN HONG KONG
ERNST & YOUNG

From the supplier's perspective, knowledge of Hong Kong's tax regime is important at two fundamental levels: What are the tax consequences of choosing intermediaries who sell on one's behalf, such as agents or manufacturer's representatives, as opposed to a distributor who buys on his own account? And what are the tax consequences for suppliers who expand their role by establishing sales branches in Hong Kong? These questions will be addressed in the following overview.

PRINCIPAL TAXES

Introduction

Companies operating in or from Hong Kong enjoy many tax advantages. Hong Kong does not have a comprehensive system of taxation; instead it has separate schedules for the taxation of different types of income arising or derived from Hong Kong. Capital gains and dividends are not taxed, and all offshore income is outside the scope of Hong Kong tax.

Direct taxes

There are three direct taxes in Hong Kong: a profits tax on business profits, a salaries tax on pensions and income from employment, and a property tax on income from real estate property.

Profits tax is imposed on every person carrying on business in Hong Kong for assessable profits derived from Hong Kong. It is not imposed on profits from the disposal of capital assets. Corporations, branches, partnerships, joint ventures, trusts, individuals and unincorporated bodies of persons doing business in Hong Kong are charged profits tax.

A salaries tax is generally imposed on individuals who derive Hong Kong-source income from any office or employment, including all income related to services rendered in Hong Kong. Income from any office or employment includes wages, salary, commissions, gratuities, bonuses, perquisites, leave pay and pensions.

Property tax is imposed on the owners of land and buildings in Hong Kong for amounts earned from renting the land and buildings. Property occupied or used by a taxpayer for business purposes that generate income subject to profits tax may also be exempted; however, even if property tax must be paid, it can be offset against the profits tax liability.

The three taxes are separate and distinct, and each is imposed only on income from Hong Kong sources and certain income deemed to derive from Hong Kong sources. Consequently, income that does not fall under one of the above categories or does not derive from Hong Kong is not subject to taxation. Hong Kong's tax system does not have a total income concept, other than a personal assessment election available to resident individuals to aggregate various sources of income.

Direct taxes in Hong Kong account for over 60% of the total government revenue.

Indirect taxes

Hong Kong has not yet levied a general consumption tax but it has numerous indirect taxes including user fees, rates, betting duty, and various other selective taxes and charges. In addition, a stamp duty is levied on documents relating to transfers of real estate and Hong Kong stocks and shares.

Indirect taxes account for approximately 25% of total government revenue. The remaining 15% of government revenue is derived from sales of land, properties and investments, and various charges and other receipts.

Tax administration

Administration of the Inland Revenue Ordinance is entrusted to the Inland Revenue Department (IRD) which is under the Commissioner of Inland Revenue. The IRD issues tax returns and assessments based on the information filed. The assessment shows the amount assessed, the tax payable and the due date for payment. There is no self-assessment requirement but taxpayers must report their liability to the IRD to enable it to issue returns.

The fiscal year-end is March 31, but liability to profits tax is based on the results of the accounting year ending in that year of assessment rather than on profits arising during the actual fiscal year.

Assessment and appeal The IRD also has the power to make estimated assessments and to seek information from taxpayers. Under the provisions of the Inland Revenue Ordinance, provisional assessments to tax are made each year and are based on the profits of the preceding year. The amount paid is applied against the agreed final liability of the year in question, and any excess is applied against the provisional tax payable for the succeeding year.

In the initial year of assessment, the provisional profits tax assessment is based on the results of the first six months. This figure is used as a base, pending finalization of the financial accounts of the first period. Assessments for provisional salaries and property tax may be based on estimates provided by the taxpayers who have the right to object to assessments and to appeal to the Board of Review, the first appellate review body in Hong Kong's tax administration, or to the courts.

Payment There are no prescribed due dates for payment of tax; taxpayers are notified of the due dates in their own assessments.

Tax audits There is no concept of tax audit other than in cases of suspected fraud or evasion. Tax returns are subject annually to a question-

and-answer procedure which may or may not lead to tax adjustments, possibly through the courts.

Penalties Taxes not paid on the prescribed due dates will be subject to an immediate 5% surcharge. The penalty is increased to 10% if the failure extends for more than six months.

The IRD may also agree to defer payment of taxes that are the subject of dispute; however, any taxes assessed and deferred will be subject to an interest charge if eventually determined to be payable.

There are no penalties resulting from any tax adjustments eventually determined. Late filing of returns, however, can result in penalties of up to three times the tax due on the return.

Statute of limitations The statute of limitations is six years for normal assessment proceedings. In cases of fraud or evasion, there is in practice no time limit for assessment.

RESIDENT CORPORATIONS

Rates

The current rate of profits tax for limited liability companies is 17.5%.

Determination of taxable income

Territoriality Corporations carrying on a trade, profession or business in Hong Kong are subject to profits tax on income that arises in or is derived from Hong Kong. Income that does not arise in or is derived

from Hong Kong is not subject to taxation in Hong Kong. To determine the source of a particular profit, it is necessary to look at all the relevant underlying facts and circumstances. After considering this question, the courts developed the concept of the originating cause of the profits earned (sometimes called the operations test). For trading operations, the places where orders are accepted and where sales contracts are negotiated and concluded are of critical importance.

Whether commissions are subject to profits tax depends on where the services are rendered. If no services are rendered, as in the case of territorial infringement commissions, the Inland Revenue may look to the place where any relevant agreement was signed to determine the source of the income.

No single factor or set of factors will determine the source of profits. The matter requires a thorough examination of all relevant facts.

Income The taxable profit is based on the profit shown in the audited financial statements of the business, subject to adjustments provided by specific provisions of the Inland Revenue Ordinance. All incorporated companies in Hong Kong are required, under the Hong Kong Companies' Ordinance, to prepare audited financial statements for each fiscal year.

Dividends received are excluded from taxable profit. In addition, they are not subject to withholding tax when paid. Interest and royalties from Hong Kong sources are subject to tax as normal income.

It is not possible to apportion a particular profit for Hong Kong tax purposes. However, if a corporation has income from both Hong Kong and non-Hong Kong sources, it is only the Hong Kong element that is taxed, with a corresponding proportional disallowance of the expenses incurred.

Capital gains Capital gains are not subject to tax, and capital losses may not be included in the computation of profits tax.

Deductions All expenses incurred for the production of assessable profits are allowed as deductions for profits tax purposes. Expenses such as the following are allowable:

- Rent and rates on buildings occupied for the purpose of producing profits.
- Employee remuneration and allowances.
- Repairs and replacements of equipment used in the production of profits.
- Bad debts written off and provisions for specific doubtful debts, to the extent that they have been previously included as trading receipts.
- Interest and related expenses on funds borrowed for the purpose of producing taxable profits (legislation limits deductibility to interest paid to recipients that are subject to tax in Hong Kong or are overseas financial institutions).

The following items are non-deductible for tax purposes:

- Depreciation provided in the accounts other than at rates prescribed in the statute.
- The cost of improvements.
- General provisions for bad debts.
- Expenditure of a capital nature.
- Interest paid under back-to-back loans or other contrived financing arrangements.

Depreciation allowances After the tax-adjusted profits have been determined, depreciation allowances are computed on certain capital expenditure incurred to carry on a trade or business in Hong Kong. In addition to an annual allowance, some capital expenditure also qualifies for an initial allowance which is the stated percentage of the initial capital expenditure. The initial and annual allowances vary depending on the nature of the expenditure.

Plant and machinery qualify for an initial allowance of 60% in the year of purchase. Depending on the type of plant or machinery acquired, these assets are also granted an annual allowance of 10%, 20% or 30% of the reducing value (original cost less initial and annual allowances). A pooling system is used to determine annual allowances for plant and machinery. The assets are divided into three separate pools for each of the annual allowance rates of 10%, 20% and 30%. The balance of a pool is increased for purchases and decreased for initial and annual allowances and proceeds on disposal.

Industrial buildings used in carrying out a qualifying trade are entitled to an initial allowance of 20% together with an annual allowance of 4%. Both allowances are based on the cost of construction, excluding the cost of land.

Commercial buildings, other than industrial buildings that are used to carry on trade, are granted an annual rebuilding allowance of 2% of the capital expenditure incurred in the construction of the commercial building.

Losses

Corporate tax losses are deductible from other taxable profits arising in the year of assessment, and any unrelieved losses are carried forward indefinitely and applied against any future assessable profits. Losses may not be carried back nor be relieved against profits of a parent, subsidiary or fellow subsidiary company. Changes in ownership or in the nature of the trade of the company do not affect the company's right to carry tax losses forward for offset against future profits.

In general, tax loss carryforwards are not forfeited when a loss company is acquired by another entity. Provisions exist, however, to prevent a profitable company from acquiring a firm with tax losses to obtain the benefit of the losses by the transfer of a profitable operation, and hence income, to such a company. If the sole or dominant purpose of an acquisition is tax avoidance, then the losses may be disallowed. The legislation refers to any change of shareholding. Care must therefore be taken in structuring group reorganizations and reconstructions. Advance tax rulings can be obtained if necessary.

NON-RESIDENT COMPANIES

Taxation of branch operations

Foreign corporations are taxed to the extent that they have Hong Kong-source profits from a trade or business carried on in Hong Kong. The corporation, rather than the Hong Kong branch, is the legal entity subject to Hong Kong taxation. The factor governing whether the corporation is subject to tax is the extent to which the corporation has profits arising in or derived from Hong Kong.

The assessable profits are generally determined by the financial accounts of the local establishment. If the Inland Revenue Department does no accept the financial accounts as reflecting the true profits of the establishment, one of the following methods of determining the taxable profits may be used:

- $\dfrac{\text{Hong Kong income}}{\text{World income}}$ X World profits (as adjusted for Hong Kong tax purposes)
- a fair percentage of the turnover of the business in Hong Kong

Imports

A foreign corporation making direct sales to Hong Kong must determine whether it is trading with, or trading in, Hong Kong. If the foreign corporation is trading with Hong Kong, its profits are non-taxable. If it is trading in Hong Kong and has profits from its Hong Kong operations, it is liable to Hong Kong taxation.

Profits on sales contracts negotiated and concluded outside Hong Kong should not attract Hong Kong taxation. If, however, sales are conducted through an agent in Hong Kong who negotiates and concludes the contracts there, the agent will be taxed on his or her commission from these transactions, and the foreign corporation will be liable to tax in Hong Kong because it was trading through an agent in Hong Kong. In contrast, if the agent's activity in Hong Kong is restricted to merely soliciting orders that are subject to acceptance or rejection by the foreign corporation outside Hong Kong, the profits accruing to the foreign corporation would not be liable to Hong Kong tax.

Reference is often made in Hong Kong to a 0.5% consignment tax payable on the gross proceeds of sales in Hong Kong on behalf of non-residents. Although there is no consignment tax, a person selling goods in Hong Kong on behalf of a non-resident may be required to pay 1% of the proceeds to the Inland Revenue Department. The Inland Revenue Ordinance does not indicate the precise nature of this amount; nevertheless, it appears to be a payment on account of the profits tax liability of the non-resident on its profits from doing business in Hong Kong.

A foreign corporation that has paid the 1% through its agent may apply to be assessed for profits tax, as a *quasi* branch, with a credit or refund for the amount deducted by the agent and paid over to the Inland Revenue Department.

Representative in Hong Kong

A foreign corporation that restricts its operations in Hong Kong to representative functions will not be taxable in Hong Kong because it will not be carrying on business in Hong Kong.

INDIRECT TAXES

Stamp duty

Stamp duty has limited application and is levied only on assignments of immovable property, leases and share contract notes and transfers. The rate of duty on property transfers is 2.75% per HK$1,000 on transfers exceeding HK$1,500,000. The rate of duty on share transfers is HK$4 per HK$1,000, payable one-half by the purchaser and one-half by the seller.

Capital duty

Capital duty is payable on the authorized capital of a company, regardless of whether it is issued, at the rate of HK$6 per HK$1,000.

Business registration tax

Every business carried on in Hong Kong is required to be registered with the business registration office. The annual registration fee is HK$1000, plus an additional levy of HK$150 for a fund to pay the wages of redundant employees of insolvent businesses.

Rental rates

Rates are levied on land and buildings on the bases of annual rental value which is calculated by reference to rent earned or deemed to be earned.

Import and excise duty

Hong Kong is a free port and has no general tariff on imported goods. Alcoholic beverages, tobacco, certain hydrocarbons and methyl alcohol are subject to duty, however, if they are imported into or manufactured in Hong Kong for local consumption. Perfumes and cosmetics are also subject to import and excise duty.

Automobiles and trucks are not subject to duties but a preliminary registration tax must be paid to the Transportation Department.

Hotel accommodation tax

A 5% hotel accommodation tax is imposed on accommodation charges paid by guests.

Entertainment tax

An entertainment tax is imposed on the price of admission to race-tracks at rates that vary depending on the admission charge. Effective April 1992, the entertainment tax on cinema tickets was abolished.

Airport tax

All persons over 12 years of age leaving Hong Kong by air are charged an airport tax of HK$150. The rate for children 2 to 12 years of age is HK$50. Children under two years of age are exempt from the tax.

Taxes not imposed in Hong Kong

The following taxes are not imposed in Hong Kong:

- Withholding tax on dividends or repatriation of profits.
- Sales or value-added tax.
- Payroll tax.
- Capital gains tax.

TAX TREATIES

Hong Kong has no tax treaties with other countries, and profits from Hong Kong sources are taxable in Hong Kong regardless of whether overseas tax has been paid. Limited tax credit relief is available for taxes payable in some Commonwealth countries other than the United Kingdom. Although not creditable, foreign taxes payable on Hong Kong taxable income may, in certain circumstances, be deductible.

A reciprocal shipping agreement with the United States became effective in 1989.

FINANCIAL REPORTING AND AUDITING IN HONG KONG

Statutory requirements

A company incorporated under the Hong Kong Companies Ordinance is required to maintain the following statutory books and registers at its registered office in Hong Kong or at another office in Hong Kong where they are maintained:

- Register of members that includes the names, addresses and occupations of members, their shareholding, amounts paid on the shares and the date at which a person becomes or ceases to be a member.

- Register of directors and secretaries that includes the names, addresses, nationalities, occupations and Hong Kong identity card numbers of directors and secretaries.

- Register of charges that records the assets of a company which have been pledged as security.

- Books containing the minutes of general meetings and meetings of the company's directors.

- Register of holders of debentures or debenture stock. This applies only to a company that issues debentures or debenture stock not transferable by delivery.

Companies incorporated under the Companies Ordinance are required to maintain proper books of account. Such books must list all assets and liabilities, as well as all receipts and expenditures, with proper explanation and support for all items and an account of all purchases and sales. The books may be kept either at the registered office of the company or at another location chosen by the directors. If the books are maintained outside Hong Kong, sufficient records must be maintained in Hong Kong to give a reasonably accurate portrayal of the company's financial position. A company's books of account must be preserved by the company for seven years from the end of the financial year in which the last entry was made or matter was recorded. Companies are permitted to maintain accounts in any currency.

The company's directors are required to present a balance sheet and a profit and loss account every calendar year at the members' annual general meeting. A private company (not a member of a group of companies that includes a listed company) may have a company as a director. If it has subsidiaries, group accounts must be presented. The accounts must cover a period ending not more than six months prior to the date of the annual general meeting.

SOURCES OF ACCOUNTING PRINCIPLES

Accounting standards and guidelines of the Hong Kong Society of Accountants

The principal source of accounting principles in Hong Kong is a series of accounting standards and guidelines issued by the Hong Kong Society of Accountants and distributed to its members. These standards and guidelines have no legal force but derive their authority from the Society which may take disciplinary action against any of its members responsible, whether as preparer or as auditor, for accounts that do not follow the requirements of the pronouncements. Dealing with matters of accounting measurement as well as disclosure, the standards and guidelines are regarded as strongly persuasive in interpreting the legal requirement that accounts should give a true and fair view.

Because a significant number of the Society's members were trained in England, Scotland and Australia, the accounting practices and audit procedures adopted in Hong Kong tend to reflect the influence of the accounting bodies of those countries. The accounting standards and guidelines issued by the Society are generally based on statements issued by the Institute of Chartered Accountants in England and Wales.

The accounting standards and guidelines are not confined in their application to incorporated companies but apply to all entities that prepare accounts intended to give a true and fair view, unless specifically exempt by a particular pronouncement. In addition, companies may depart from them if they are in conflict with disclosure exemptions granted by law, such as certain Companies Ordinance exemptions afforded to banks and insurance companies.

Legislation

The Hong Kong Companies Ordinance confines itself to matters of disclosure rather than development of rules on accounting measurement which has been left to the Hong Kong Society of Accountants.

International accounting standards

The Hong Kong Society of Accountants is a member of the International Accounting Standards Committee (IASC). Accordingly, in the public interest, the Society has undertaken to support the IASC's objectives of formulating and publishing basic standards to be observed in the presentation of audited accounts and financial statements and to promote their worldwide acceptance.

The Society has established the following guidelines for its members in connection with the IASC standards issued and in force:

- If the Society's pronouncement agrees with or goes further than the IASC standards, compliance with the former ensures compliance with the latter.

- If an IASC standard is in effect but the Society has not issued a pronouncement on that subject, the IASC standard is not obligatory in Hong Kong unless some other consideration applies.

- If an IASC standard goes further than a pronouncement of the Society, compliance with the latter is required. Non-compliance with the IASC standard would generally be noted.

Significant Hong Kong accounting principles and practices

Group accounts A company with one or more subsidiaries must present its accounts in the form of group accounts (which normally means consolidated accounts) unless it is itself a wholly-owned subsidiary of another Hong Kong company. A company is deemed to be a subsidiary of another company if it controls the composition of its board of directors, controls more than half of its voting power or holds more than half of its issued share capital. A subsidiary may be excluded from consolidation if the company's directors are of the opinion that

- it is impracticable, or would be of no real value to members of the company, in view of the insignificant amount involved, or would involve expense or delay out of proportion to the value to members of the company.
- the result would be misleading or be harmful to the business of the company or any of its subsidiaries.
- the business of the holding company and that of the subsidiary are so different that they cannot reasonably be treated as a single undertaking.

Investments in subsidiaries not consolidated are generally accounted for on an equity basis if the results are

material; otherwise, the investments are accounted for at cost.

New subsidiaries are generally accounted for by the purchase method. Merger accounting is rare in Hong Kong.

Associated companies An associated company is one that is not a subsidiary of the investing company but in which the investing company has a long-term investment amounting to 20% or more of the investee's equity shares and over whose commercial and financial policies the investing company is in a position to exercise significant influence. Hong Kong accounting principles require the use of the equity method of accounting for such investments.

Foreign-currency translation Normally each asset, liability, revenue or cost arising from a transaction denominated in a foreign currency should be translated into the reporting currency at the exchange rate in effect on the date on which such a transaction occurred. If the rates do not fluctuate significantly, an average rate for a period may be used.

In preparing group accounts, the closing-rate method of translating the financial statements of the foreign subsidiaries should normally be used. Any resulting exchange difference should be recorded as a movement on reserve.

Deferred taxes Statement of Standard Accounting Practice (SSAP) No. 12 of the Hong Kong Society requires that deferred taxes be computed under the liability method to the extent that it is probable that a liability or an asset will crystalize. In practice, it is not common to recognize deferred assets.

Depreciation Fixed assets with finite lives must be systematically depreciated to their estimated residual values over their expected useful lives so as to allocate the cost as fairly as possible to the periods expected to benefit from their use. No particular method is mandatory but the straight-line and declining-balance methods are the most common. If an asset has been revalued, depreciation should be based on the valuation rather than the cost. When the book value of an asset is not recoverable in full from future revenues, it must be written down immediately to its recoverable amount.

Inventories Inventories (stock) must be carried at the lower of cost or net realizable value. Unless specific identification is practicable, cost is generally determined on a first-in, first-out (FIFO) or an average basis. The use of last-in, first-out (LIFO) and base stock is not permitted.

Research and development costs There is no official pronouncement in Hong Kong governing the accounting for research and development costs. Research costs are usually expensed as incurred. Development expenditures are also usually expensed but they may be deferred to future periods if they can be identified as relating to a clearly defined and viable project and if recovery of such expenditure against future revenues from the project is reasonably assured. Any expenditure deferred in this way is amortized on a systematic basis against these future revenues or written off whenever it is recognized as non-recoverable.

Leases Accounting for leases is governed by the provisions of SSAP No. 14. Leases are classified as either operating leases or financial leases. In an operating lease, only the rental is taken into account by the lessee. A lease is presumed to be a financial lease if the present value of the minimum lease payments equals 90% or more of the fair value of the leased asset at the inception of the lease. A financial lease must be capitalized by the lessee together with simultaneous recognition of the obligation to make future payments.

Related-party transactions There are no specific disclosure requirements other than the need to show balances with other group companies at the end of the year and to reveal the identity of the ultimate holding company. Some take the position that such transactions must be shown in order to satisfy the requirements to present true and fair view statements, but, in practice, detailed disclosures are uncommon.

Earnings per share Listed companies are required to show earnings per share based on profits after taxation but before extraordinary items. Diluted earnings-per-share figures must also be shown if the effect of dilution exceeds 5%.

Current-cost accounting Current-cost accounting is not normally practised in Hong Kong.

Financial reporting

The disclosure requirements of statutory accounts are detailed in the Companies Ordinance and are broadly similar to those of the United Kingdom.

The Companies Ordinance requires that accounts contain a balance sheet and a profit and loss statement supported by a variety of notes. SSAP No. 4 of the Hong Kong Society requires all but the smallest entities to also include a statement of changes in financial position. Corresponding figures for the previous financial year must be presented as well.

At present, the profit and loss statement need not be a comprehensive accounting for all items of revenue and expenses included in arriving at profit before taxation; the Companies Ordinance requires disclosure only of selected items.

All significant accounting policies of the reporting company must be disclosed, and these are usually grouped together in one note or in a separate statement. Other notes usually presented include details of the movements of fixed assets, share capital, and reserves; analyses of loans payable and inventory balances; disclosures of commitments and contingencies; and details of various categories of expense such as directors' remuneration, auditor's remuneration, depreciation and taxation. The notes form part of the accounts and are included in the scope of the audit opinion.

A copy of financial statements, together with a copy of the directors' report and the auditor's report, are required to be sent to each shareholder at least 21 days before the annual general meeting of shareholders.

Within 42 days after the annual general meeting, all companies are required to file with the Registrar of Companies an annual return containing the meeting date, details of the capital structure of the company, a list of amounts due on mortgages, information regarding shareholders, a list of directors and secretaries, and all business names under which the company carries on business. The annual return of public companies must be accompanied by a signed copy of the audited financial statements and directors' report.

MANAGING
YOUR DISTRIBUTION
RELATIONSHIPS

LAUNCHING YOUR PRODUCT

A good product can be strangled at birth, even in the hands of a capable distributor, if the supplier is under the illusion that sales somehow take off spontaneously. A good launching program and effective training are absolutely necessary in a well-organized exporting drive.

OUT OF SIGHT...

The starting pistol

One of the most common misconceptions among exporters is that a swarm of salesmen will crisscross Hong Kong showing their products and booking orders as soon as the distribution agreement is signed. They could not be more wrong!

The day after the deal is signed, the distribution company owner has probably had time to think of ten good reasons why your product is no good. The product manager may have chucked your product literature in the filing cabinet because he has more pressing problems to deal with. As for the salesmen, they are probably unwilling to demonstrate your product because they only remember 5% of what you told them and do not want to appear foolish in front of prospective buyers. If thousands of exporters face such situations every year, it is because they mistake a signature for a commitment when all the distributor wished to do is purchase an option for himself.

What is needed at this point is a well-executed launching program. While this program should be mutually agreed to before the distribution

agreement is signed, you, the supplier, must realize that commitment to your product will not flow automatically from a contract and cannot be expected as a right. Commitment is based on human relations and concrete results, both of which can be developed by a conscious effort. This is where the field rep's personality and communication skills really come into play.

Russian roulette

The lack of time is usually cited as the reason so many exporters fall short at this crucial point. Exporters typically swing into Hong Kong to exhort their distributors over drinks, then move on to Seoul or Taipei for more pep talks over more drinks, followed by Singapore, Bangkok, Sydney, Melbourne, etc., where the same scene is repeated. Many exporters believe that if ten wet wicks are lighted, one is surely bound to go off! The problem is not time; plenty of it is wasted. Export managers simply must realize that distributors in Asia will not perform well unless field reps have the time and skills required to launch the product properly. Once you have prioritized your target markets, it is far more profitable to develop them one by one in a thorough fashion than to try to cover all of Asia in one fell swoop.

BACK TO THE PLAN

Making goals explicit

Launching programs are usually settled during the final stages of the contract negotiation process. Your goal at that stage is to take the distributor's senior management out of the realm of theory by committing them publicly to subsequent planning meetings, press releases, sales training sessions, etc.

Sales targets

This program is part of a larger sales promotion plan which must also be made explicit. Amateurs often resort to pleading and bullying to push sales; experts rely on negotiated plans to ratchet up distributor performance and develop their in-house market knowledge of Hong Kong and the PRC. The plan need not be perfect or even good at first: as long as it is reasonable and gives the distributor clear and credible sales targets for the coming year, your product should attract its fair share of distributor resources during the first crucial twelve months when its fate will be decided. The two most current methods used to define sales targets in the early stages are the market factor method and the promotion plan approach. In both cases, your persuasiveness in the eyes of the distributor will depend on your familiarity with

competing products and the customer's buying habits (see Getting the Big Picture).

Market factor method

To avoid haggling over numbers until both sides reach the midpoint between two extreme positions, certain suppliers try to identify an independent variable and a target market share to which sales targets can be tied. This independent variable – called market factor by marketing specialists – should have a causal relationship with the level of demand for your product.

Example

If a baby crib manufacturer agrees with his distributor that the relevant market factor is the number of births during the previous year and that 5% is a reasonable market share to aim for during the first year, the sales target could be calculated as follows:

Known live births	
in Hong Kong	
(1991)	*68,508*
times market share	*x .05*
Sales forecast for	
1992	*3,425*

Although market factors are not always as easy to determine as in our example, the advantages of this method are its objectivity and face validity. The distributor can follow its logic without a knowledge of statistics and marketing research. However, this method still leaves plenty of room for quibbling. A distributor could counter that most apartments are too small for cribs. More importantly, the supplier has not nailed down the distributor as to *how* he intends to sell that number of cribs.

Promotion plan method

Because distributors often think in terms of man-hours spent to achieve certain returns rather than market share, certain suppliers think it best to first reach an agreement on a promotional plan before attempting to forecast sales. This is the approach we also recommend. If you have done your research properly, you know something about competing products, who your target customers are and through which intermediaries they are best approached. The next step is to prioritize your customer groups with your distributor according to sales potential and list the various ways which can be used (along with their cost and expected results within a year) to stimulate sales for each of these groups. What will ensue is a classical bargaining session where a balance is struck between what you would like and are willing to invest in the sales effort, and what the distributor is willing to contribute. Then, and only then, agree to sales targets.

The advantage of this system is that the distributor has a closer stake in achieving his forecasts and a stronger sense for the ways in which the product can be made to grow. If the first year is a success, you can often incorporate the market factor line of reasoning in your argument for augmenting sales targets. By then, the distributor will be far more trusting and receptive.

The product specialist

One final and crucial point to emphasize in a promotion plan is the principle of having a product manager or senior salesman designated for special training related to your product. Ideally, this should be someone with whom you communicate easily and one with some authority and seniority (remember Hong Kong's turnover rate). This point cannot be overemphasized. Productive franchises often owe their success to the fact that someone within the distributor's organization became personally and enthusiastically involved with the product. As "product specialist", he will be the one who deals with key and demanding customers, keeps the sales force motivated and maintains inventory levels in your absence. Insist on a product specialist regardless of what the distribution company owner says. If there is no way of getting one, seriously consider giving your business to someone else.

KICKOFF

Promotion

To reach a dispersed customer base, trade fairs and exhibitions are the usual way Hong Kong distributors announce the launching of a new product. For more highly focussed customer groups, they are willing to invest in technical seminars and mailings. Advertising is another matter, often giving rise to disagreements between suppliers and distributors. The general feeling among distributors is that anything fancy and expensive whose returns cannot be quantified should be paid for in part or in whole by the supplier.

Timing

The launching program should follow fast on the heels of the contract. Do not allow the distributor to lose enthusiasm for your product by allowing too much time to elapse between signing the agreement and concrete action.

KISS

This is no time to waste resources on fringe products and territories. Reference has already been made to Hong Kong's limited shelf space and the high cost of storage. Limit your product line during the first year to the few items most likely to succeed. Similarly, concentrate your promotional program on the territories where success is most assured. This is where your early research (see Getting the Big Picture) will help you tremendously to steer initial efforts most productively. More difficult territories can be tackled later when you have momentum.

Concessions

Expect your distributor to ask for price concessions, even during the launch. They do this to test you and to

make the point that no price list is sacrosanct. It is very important that you resist any such demands or you will never stop being badgered for more discounts. Moreover, new products are, by their very nature, self-promoting and it is unwise to squander special discounts during the launch.

Translation

Another occasional bone of contention is the translation of sales material. English is usually acceptable for technical products not meant for home use. Consumer products, on the other hand, require Chinese translation. For certain product categories, too, translation on the packaging helps prevent parallel imports (see Weak Points and Trouble Spots).

TRAINING

Crucial

Training is probably the most powerful tool in the supplier's hand to win the war for distributor attention. It is a basic, ongoing necessity for success with distributors, and never more important than at the beginning. If done well, it will go a long way towards enhancing the confidence of the sales force in your product and simplify communication when problems arise. Done in a careless, lackluster way, it will almost certainly lay your product to rest at the bottom of everyone's priority list.

Presentation

Training presentations work best if certain fundamental rules are followed:

- Assume tactfully that your audience knows nothing.
- Speak slowly and avoid slang expressions, especially if your speech is being interpreted.
- Keep technical questions for a separate session.
- Tailor your message to the special interests of your audience: senior management, your product specialist, the sales force, technicians (see section below).
- Pay proper homage to any superiors who may be present by mentioning their names and soliciting comments.
- To inject some excitement into the presentation, use visual aids, maintain some physical movement and avoid reading a prepared script.
- Move from the general to the specific. After a brief, self-deprecating and informal introduction, give a general industry background as it relates to Hong Kong and the PRC. A general product background which highlights customer needs left unsatisfied by the competition is useful before you present your own product again in terms of the customer needs already identified. Provide a market synopsis outlining the prioritized list of customer groups previously identified with distributor senior

management. The salesmen will be most interested in your view as to how these customer groups can best be reached. Finally, summarize the key points covered.

Launching

The training sessions provided at the beginning of the launching process should incorporate all of the elements identified above. A special effort must neverthless be made to keep your presentation to the point and enjoyable. Limit yourself only to the few items they will actually be selling during the initial months; emphasize market training and selling skills; and make sure there are plenty of opportunities to establish one-on-one relationships with the sales force.

Failure

Sadly, many North American field reps do poorly in training presentations for three reasons: they do not know their product sufficiently to answer questions from what can be a sophisticated, internationally-minded audience; their knowledge of the Hong Kong/PRC market is insufficient to capably prioritize different customers groups and give meaningful advice as to how best to reach them; and they underestimate the importance of communication skills and personal relationships. Family-owned distributing firms, in particular, respond very well to a form of guidance that is not overbearing yet demonstrates a sincere interest in their well-being (what some would call paternalism).

TRAINING DIFFERENT AUDIENCES

Owners/senior managers

Assuming the field rep has done his original market prospecting thoroughly (see Getting the Big Picture), much of the training at this level will occur during the process of formal product presentations and negotiations. Subsequent training meetings should emphasize your product in a competitive perspective, its market potential, its compatibility with the distributor's other lines and, of course, its profitability.

Product specialist

This is the person, mentioned earlier in this section, selected for special training because you require someone within the distributor organization who knows your product almost as well as you do. The other (unavowed) goal is to establish a personal relationship with that person so that he becomes committed to your success. The most time-effective way to reach these objectives is usually to invite the product specialist to your home plant. He will not only gain a first-hand impression of your capabilities but you also have his undivided attention and a much greater chance of winning his dedication and loyalty because overseas trips are a special

event, long to be remembered. Make sure the distributor has a stake in the product specialist's training. The usual way to share costs is for the distributor to pay for travel expenses while you pick up on-site costs, such as room, board and incidentals.

Salesmen and technical personnel

Because the salesman's greatest challenge is the fear of rejection, arising especially from an ignorance of the product and its application, the most effective training at this level is done through small group sessions followed by hands-on field sales training. One hour of field work is worth a hundred hours of lecturing. Almost as important are the many opportunities field work gives you to strike up a personal relationship with various salespeople. Such relationships go a long way in a Chinese context towards ensuring that your product gets more than its fair share of attention after you are gone.

Sales types and technicians rarely mix well together. It is best, therefore, to organize different sessions for both groups, making sure that the presentation for the technicians is given by a qualified person able to provide all the answers required. Complete familiarity with the product's technical and maintenance requirements is the only way to gain the respect of the engineering and technical staff.

MOTIVATION

The best promotional plans involve incentives for all the crucial players in a sales effort: the distribution company owner, his sales staff and the ultimate buyer. Attention must be paid to all of these groups.

MOTIVATING THE DISTRIBUTOR

Four tools commonly used to motivate distributors are pricing/margins, training, internal promotion and overstocking.

Pricing/margins

Thousands of books have been written on the subject of pricing. We will limit ourselves to five basic principles of particular relevance to Hong Kong:

- First and foremost, a unique product, effective training, an enthusiastic sales force, high sales volumes, stunning product literature – none of these is worth anything if, at the end of the day, both supplier and distributor, are not making money.

- Even if money is being made, the relationship will be unstable if one side is making a 50% gross margin while the other is cutting profits to the bone in order to maintain the product's competitiveness. If both sides are interested in a stable relationship, the profits will have to be balanced for both partners.

- In a fast-moving business environment such as Hong Kong's, this balance cannot be reached through a rigid application of legalistic formulas. Pricing and margins involve a continuous process of give and take within guidelines agreed to in the sales promotion plan. For example, a distributor who sacrifices profits because of an upswing in the purchase currency might be compensated for a period with better margins when the situation has corrected itself. Knowing when to give some slack or rein in margins to maintain competitiveness is all part of the distribution game. To be successful at it, suppliers have to be involved, do their calculations and stay informed.

- Distribution in the PRC often entails exceptions to this principle of balance because certain intermediaries may seem to be making large profits. It must be remembered, however, that this money must be shared with a host of contacts and that competition there, too, is very stiff. Controlling the product's ultimate selling price is not always possible or desirable. Its competitiveness within the target market is all that counts.

- It is always a wise precaution to structure your pricing to allow something extra for promotional activities even if these activities may have already been promised by the distributor in the joint sales promotion plan (see Launching Your Product). With so many suppliers to satisfy, it often happens that a distributor's promises fall through the cracks unless specific incentives (e.g., price reductions, free trial samples, cost sharing, cash bonuses for the sales team, etc.) are tied to them. Arguing that the distributor's margin is already sufficient may win you the battle but you may ultimately lose the war. Accept these realities of Asian distribution and plan for it from Day One.

Training

By simplifying communications and adding to the sale staff's self-confidence, training can be a powerful motivating force for a distributor. To be effective, however, especially with Hong Kong's high turnover rate, training has to be treated by the supplier not as a one-shot deal during the launch but as part of an ongoing process with performance evaluations, check points, and milestones.

Internal promotion

Large Hong Kong distributors accept cash bonuses for sales staff as long as they are paid out to teams rather than individuals. Prizes for individual salesmen tend to be of the symbolic kind. Smaller distributors may object to internal promotion because they find them disruptive. Where they are allowed, internal promotions work best when they are fair, simple to understand, and tied to specific targets.

Overstocking

Inventory is not usually thought of as a motivator; excessive inventory levels force distributors to put pressure on their sales staff to off-load products at a faster rate. Suppliers, therefore, regard overstocking as a legitimate pressure tool as long as their distribution agreements have a no return-of-merchandise clause. The methods used to incite distributors to overstock include persuasion, bullying, anticipated price increases and real or imaginary threats to supply. Well-organized distributors who adhere to joint sales promotion plans rarely fall for this ploy but many smaller firms do. It is worth noting that overstocking is a well-known "push strategy" commonly used by Hong Kong distributors with poorly organized PRC wholesalers and retailers (see Market Entry and the Hong Kong Middleman).

MOTIVATING THE CUSTOMER

How to stimulate the end-user to purchase your product is a vast field already well served in many marketing publications. We will limit ourselves to four general principles which should help you make effective use of distributors in Hong Kong.

The plan again

The anchor and guiding light for a productive relationship between a supplier and his distributor is the joint annual sales promotion plan (see Launching Your Product). Every day, in hundreds of offices in Hong Kong, hours are wasted in pointless arguments over sales figures because the parties concerned never set objectives or made tradeoffs between what the supplier wanted and what the distributor was willing to invest.

Simplify

Hong Kong distributors appreciate sales programs which involve media where goals and return on investment can be quantified and measured. Anything more, such as television advertising, will require partial, or more likely, complete funding from the supplier.

Sales campaigns should also be simple. It is better to have three well-organized campaigns than six half-hearted efforts from an exhausted sales force. For an expensive market, such as the PRC where prompt feedback saves resources, it is better to launch a pilot sales program in a part of Guangdong Province than grandiose sales campaigns over all of South China.

Self reliance

Never assume that the distributor, especially small ones, know anything about promotion.

Matching funds

Except for such frills as TV advertising, always make sure that the promotional program is jointly funded on a matching basis. The principle here is simple: the distributor is far more likely to implement the program thoroughly if he is paying for it than if it comes to him free of charge.

COMMUNICATION, EVALUATION AND GRIEVANCES

The rule of thumb here is that an ounce of prevention is worth a pound of cure. Indeed, the high level of support Asian distributors sometimes require allows well-organized suppliers to incorporate supervision and evaluation activities in their support programs.

COMMUNICATION

Launching

Contract negotiation is usually the responsibility of senior marketing managers on both sides. Once this is completed, it is best for them to delegate daily communications to certain subordinates and only touch base when they must deal with serious problems or make important decisions. Part of the launching process, therefore, involves the identification of which person talks to whom for the purposes of ordering, shipping, payments, technical advice, etc. Changes in personnel which affect these lines of communication should promptly be reported.

Reports

This must be defined precisely in writing during the contract negotiation process and not later because distributors hate to write reports. They rightly feel that a greater amount of information in the supplier's hands means more criticism of their performance and more unsolicited advice. Suppliers, nevertheless, need to know what is happening and those whose training programs are considered most effective are very creative in devising uncomplicated ways to routinize the distributor's information-gathering and reporting system. Critical data are usually reported on a monthly basis; information on the activities of competitors or on market trends can be scheduled quarterly.

Distributors must also be made to understand that you may occasionally require special reports for annual reviews or when evaluating the marketability of a new product. Most distributors understand these requirements. You must understand, however, that market information is sometimes very difficult to get in Hong Kong. Do not request data unless you have already discussed the availability of this information with the distributor. Distribution companies are not, after all, market research firms. Small firms, in particular, may not be able to produce sophisticated market studies.

DISTRIBUTOR'S LAMENT

Depending on the product and the nationality, the most common complaints about North American suppliers among Hong Kong distributors revolve around their attitude toward exporting and towards distributors, in general, and the behavior of certain irresponsible representatives.

Not export-minded

Hong Kong distributors feel that many North American suppliers are only interested in their own domestic market. Exporting outside North America is seen as a stop-gap measure to ensure that plant capacity is fully utilized when domestic demand is insufficient. As a consequence, exporting is not treated with the same professionalism as in Europe: market knowledge is often sophomoric; answers to faxed inquiries are slow to come and often incomplete; shipping delays are not taken seriously; and packaging is not downsized or labelled in Chinese for the convenience of the Hong Kong or PRC consumer.

Uncooperative

Another typical complaint is that North American suppliers are not used to dealing with distributors, distrust them, and do not consult with them when making decisions which may directly affect the value of the franchise, such as changes in sourcing patterns, pricing, product line composition and in ingredients or components. Worse still, certain suppliers assign sales quotas without consultation and then switch to other distributors at the drop of a hat if the quota is not reached.

Junketing firemen

All distributors, particularly the small ones, are frustrated when field reps or marketing managers make frequent visits with no defined purpose. Protocol often demands that senior people, perhaps even the owner, accompany them around and some reps have come to expect lavish meals, nights out on the town and other favors. This can become disruptive when pushed beyond a certain point.

THE SUPPLIER'S TURN

Insufficient sales

Suppliers, especially those who select distributors without studying market conditions or who assign sales quotas without a joint promotion plan, are often unhappy with their distributor's performance and are convinced things would improve if only North American marketing techniques were more extensively used.

No feedback

Many local Hong Kong distributors tend, from a Western perspective, to remain quiet about difficulties because they are unwilling to lose face in front of suppliers. This can sometimes go on until problems grow insurmountable.

The glass ceiling

Suppliers whose first year is relatively successful worry about the leveling off in growth which many products undergo when in the hands of distributors. The reason for this is simple: the distributor carries a multitude of lines; as long as each of them conforms to basic performance standards, he cares little whether his profits come from your line or someone else's. This can often lead to friction with a given supplier if the distributor is seeking an overall 5% growth rate for his business while the market for the supplier's product is growing at 15%. To resolve this, the supplier has three basic tools at his disposal:

- The most important is the annual sales promotion plan. Use it to ratchet up performance targets each year and to squeeze margins for customer groups where further sales growth is unlikely. Remember to give more generous margins to other neglected customer groups.

- The incentives which are tied to various promotion activities (see Motivation) should also be used to focus attention on new customer groups where growth potential is highest. Promotional activities should be refined to reach your target groups more efficiently than during the first year.

- Ask for a product manager for your line, preferably your "product specialist" (see Launching

Your Product) if he has proven more than satisfactory. In time, most successful suppliers also insist on a dedicated sales team. Expect some resistance to this idea because the next step for certain suppliers is to hire away this team a few years later when they decide to create their own sales office.

CANCELLING A DISTRIBUTOR

Costly

Cancelling a distributor is not something a supplier should take lightly. The risks can be real, even if they are not usually of a legal nature.

- The distributor could disrupt the market by dumping his inventory at bargain prices.

- Service levels to existing retail clients could drop precipitously, leading them to cancel your product. For technical products, the same could happen to after-sales service.

- If the distributor owes you money, collecting it or even agreeing on how much is owed could prove difficult.

- Your reputation with retailers, end-users and other potential distributors could suffer.

- In certain extreme cases, your ex-distributor could do everything in his power to make life difficult for his successor.

Controlling risks

Prevention is the best way to avoid this. Take the time and make the effort to carefully select your distributor. If problems arise after selection, do your best to ascertain the reason your distributor is not performing and help him find a solution. If cancelling is unavoidable, plan each step carefully.

- Do not cancel anyone unless you have a replacement.
- Find out how much of your product the distributor has on stock and in transit.
- If possible, avoid cancelling your distributor just before a major selling season or just after he has spent money on a major promotion program for your product.

These precautions will probably help you avoid some of the more extreme retaliatory measures we outlined above. During the subsequent negotiations, four issues in particular frequently come up:

- **Transfer of stock** Who will pay for the cost of transferring the stock from the old to the new distributor? You may have to pay part of it.

- **Trade receivables** The old distributor will want you or his replacement to pay for any stock he has sold on credit.

- **Promotion expenses** Your old distributor will want to be compensated for promotional and advertising expenses he has incurred which will only profit his successor.

- **Compensation** Some other amount may be asked for to avoid ill will or other "problems". Paying off your ex-distributor is sometime the least costly alternative.

PLANNING
AND REFERENCE

PLANNING AND REFERENCE

ENTRY DETAILS

Passports & visas

Most visitors need only a valid passport to enter Hong Kong. UK citizens (or Commonwealth citizens born in the UK or in Hong Kong) can stay up to six months without a visa. Other Commonwealth citizens do not require a visa for visits of up to three months. Depending on the country from which they originate, citizens of most Western European countries can stay for three months without a visa. Americans can stay for one month without a visa.

Visa extensions

Visa extensions are available from the Immigration Department, 2nd Floor, Wanchai Tower Two, 7 Gloucestor Rd., Wanchai, Hong Kong Island, Tel: 824-6111.

Employment/work permits

Visitor visas do not entitle you to take up employment. With few exceptions, visitors who have been offered employment must leave Hong Kong to obtain a work permit. These are made available though any British embassy or consulate.

Customs

Although Western visitors are rarely subjected to more than a few cursory questions, bottlenecks often occur at immigration; allow 60 minutes to clear the airport. Hong Kong is a free port, but this does not mean that there are no restrictions. There are the usual restrictions on firearms, explosives and drugs. As for duty-free goods, you may bring in one liter of alcohol, 200 cigarettes (one carton), 60 milliliters of perfume and 250 milliliters of toilet water.

Health

Unless you have been in an infected area in the previous fourteen days, you do not require a health certificate or any other vaccination stamp. Hong Kong's health and hygiene standards compare fully with those in the West, and travellers need take no particular precautions.

MONEY

Currency

The Hong Kong dollar is pegged to the American dollar (US$1=HK$7.80) and its banknotes come in denominations of $1000, $500, $50, $20 and $10. There are also silver coins in $5, $2 and $1 denominations, and bronze coins for 50¢, 20¢ and 10¢. Most banks are open Monday to Friday from 9:00 am to 4:30 pm, and on Saturday mornings. The banks give the best exchange rate; hotel desks, department stores and shop-front money changers charge a commission.

Credit cards

The credit cards accepted by most hotels, shops and restaurants are American Express, Bank Americard (Visa), Carte Blanche, Diners Card, JCB, Mastercard and Air Travel. Some shops may try to add a surcharge to the cost of an item if you use a credit card but this is illegal.

Travellers' cheques

European or Japanese travellers' cheques are as acceptable as US dollar cheques. Many small merchants in Hong Kong prefer traveller's cheques to credit cards or hard currency. You can therefore hold out for more than the going bank rate during any negotiations involving travellers' cheques.

Overseas remittances

The simplest, fastest and surest way to have money wired to you is through an international bank, preferably one from your home country, and ideally one where your current savings account is maintained. The easiest procedure is to have a friend or relative make a deposit in your home bank and instruct it to relay funds to a related bank branch in Hong Kong. To avoid delays, your full name, passport number and address should be included with the transfer document.

Tipping

Hotels and most major restaurants include a 10-15% service charge. Nevertheless, many waiters and hotel-cleaning staff expect more. Only give a tip if you feel that service has been exceptional. The same goes for taxis. One place where visitors are caught off-guard by requests for tips is washrooms. Lavatory janitors in major hotels and night clubs expect gratuities for good service.

GETTING AROUND

Taxis

Taxis are the most convenient means of transport for visitors. However, it is often hard to find them during rush hour, when it rains or during shift changes (around 4 pm). One of the surest ways to get a taxi is to join a line at the nearest hotel. Taxis can also be hailed in the street although there are many restrictions and taxis can only stop at certain designated areas. A yellow line drawn close to the curb indicates no stopping during morning and evening rush hours. A double yellow line indicates no stopping at any time. When unoccupied, taxis have a lighted dome at night and a red flag on the dashboard during the day.

The drop rate in Hong Kong and Kowloon in 1992 was HK$9 for the first 2km and 90¢ per .2 km thereafter. Waiting time was 90¢ per minute. Taxis are all metered. Make sure the meter is reset at the beginning of your journey. If you go through the Cross-Harbour Tunnel, you will be charged an extra HK$20; the toll is only HK$10 but the driver is allowed to double the toll because it is assumed he will not get a fare back. Drivers usually speak enough English to get you to well-known spots but if you are going out of the way, it is best to have the address written in Chinese.

Mass Transit Railway (MTR)

For short journeys across the harbor or between Causeway Bay and Central, the MTR is faster and cheaper than a taxi. Fares for the 38.6 km underground system range from HK$3-$8.50. The ticket machines take exact change but there are change booths where notes can be broken. After placing a bill in a ticket machine, you will receive a magnetized plastic card to operate the turnstiles to enter and exit the stations. There are no public washrooms in the stations and smoking, drinking and eating are prohibited. Service hours are 6 am to 1 am. For more information, call Tel. 750-0170.

Hotel transport

All the main hotels provide limousine service to their guests with English- and Japanese-speaking drivers. Charges are calculated on an hourly basis.

Driving

Traffic drives on the left, roads are very congested, and parking is difficult. It is advisable, therefore, to avoid driving a rented car in the inner city areas. The main car rental companies are Avis (Tel: 571-9237) and National (Tel: 367-1047).

Trains

Only one railway system, the Kowloon-Canton Railway (KCR), exists in Hong Kong and goes from Kowloon station in Hung Hom to the border with China. There are two main services: the daily express service to the border at Lo Wu for those going to China and the local "stopping" train which stops at every station up to Sheung Shui. Trains run every 10-15 minutes. Fares range from HK$3 to HK$7 to HK$25 for Lo Wu. The customer service hotline is Tel. 602-7799.

Light Rail Transit (LRT)

Hong Kong's light rail transit system operates in the western New Territories between the Tuen Mun ferry pier and Yuen Long. For enquiries, call the hotline at Tel. 468-7788.

Trams

On Hong Kong Island, double-decker trams run from Kennedy Town in the west, through Central to Causeway Bay and then to Shau Kei Wan in the east. Apart from a small detour some of them make around Happy Valley, they travel across the north side of the Island. The standard one-way fare which you pay on your way out is HK$1 exact fare. Take it for fun; avoid it if you are in a rush. Service hours are 6 am to 1 am, depending on routes.

Ferry

The Star Ferry operates between Kowloon and Central, and is a spectacular way to cross the harbor. The journey takes 15 minutes and costs HK$1.20 for the upper deck or HK$1 for the lower deck. Service hours are 6:30 am to 11:30 pm.

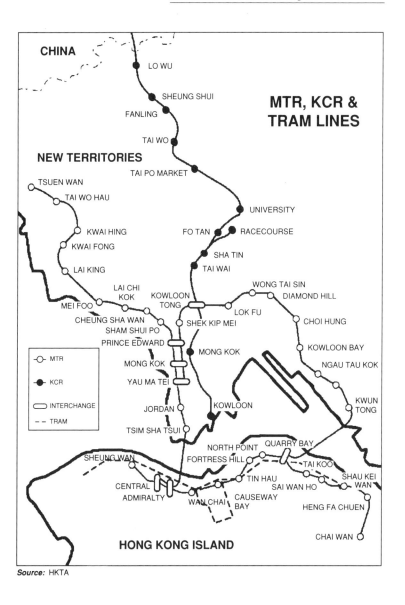

Source: HKTA

HOTELS

Hong Kong receives about 6 million visitors a year and the standard of accommodation available is very high. The better hotels are nearly always full from September to December, during Chinese New Year (late January-early February) and Easter. Suppliers visiting distributors will find it more convenient to stay on Hong Kong Island. Those visiting Guangdong or the New Territories, where inventories are normally stored, should choose a hotel on the Kowloon side. The single-occupancy rates listed below are quoted only as guidelines. A 10% service charge and a 5% room tax are charged in addition to published rates.

HONG KONG ISLAND

US$200 + (HK$1560+)

Conrad
Pacific Place
88 Queensway, HK
Tel: 521-3838
Fax: 521-3888

Grand Hyatt Hong Kong
1 Harbour Rd, Wanchai
Tel: 861-1234
Fax: 861-1677

Hong Kong Hilton
2 Queen's Rd., Central
Tel: 523-3111
Fax: 845-2520

Hong Kong Marriott
Pacific Place, Queensway, Central
Tel: 810-8366
Fax: 845-0737

Mandarin Oriental
5 Connaught Rd., Central
Tel: 522-0111
Fax: 810-6190

Victoria
200 Connaught Rd., Central
Tel: 540-7288
Fax: 858-3398

US$150-199 (HK$1170-1560)

Furama Kempinski HK
1 Connaught Rd., Central
Tel: 525-5111
Fax: 868-1768

New World Harbour View
1 Harbour Rd., Wanchai
Tel: 866-2288
Fax: 866-3388

Park Lane Radisson HK
310 Gloucester Rd., Causeway Bay
Tel: 890-3355
Fax: 576-7853

US$100-149 (HK$780-1170)

China Harbour View
189 Gloucester Rd., Wanchai
Tel: 838-2222
Fax: 838-0136

China Merchants Hotel
160-161 Connaught Rd. West,
Western
Tel: 559-6888
Fax: 559-0038

City Garden
231 Electric Rd., North Point
Tel: 887-2888
Fax: 887-1111

Eastin Valley Hotel
1A Wang Tak St., Happy Valley
Tel: 574-9922
Fax: 838-1622

Evergreen
31-39 Hennessy Rd., Wanchai
Tel: 866-9111
Fax: 861-3121

Excelsior Hotel
281 Gloucester Rd., Causeway Bay
Tel: 894-8888
Fax: 895-6459

Grand Plaza
2 Kornhill Rd., Quarry Bay
Tel: 886-0011
Fax: 886-1738

Lee Gardens
Hysan Ave., Causeway Bay
Tel: 895-3311
Fax: 576-9775

Luk Kwok
72 Gloucester Rd., Wanchai
Tel: 866-2166
Fax: 866-2622

Ramada Inn HK
61-73 Lockhart Rd., Wanchai
Tel: 861-1000
Fax: 865-6023

Wharney
61 Lockhart Rd., Wanchai
Tel: 861-1000
Fax: 865-6023

US$50-99 (HK$390-780)

China Merchants
160 Connaught Rd., West
Tel: 559-6888
Fax: 559-0038

Emerald
152 Connaught Rd., West,
Tel: 546-8111
Fax: 559-0255

Harbour
116-122 Gloucester Rd., Wanchai
Tel: 574-8211
Fax: 572-2185

Harbour View International
4 Harbour Rd., Wanchai
Tel: 520-1111
Fax: 865-6063

New Harbour
41-49 Hennessy Rd., Wanchai
Tel: 861-1166
Fax: 865-6111

Caravelle
84 Morrison Hill Rd., Happy Valley
Tel: 575-4455
Fax: 832-5881

KOWLOON

US$200+ (HK$1560+)

Kowloon Shangri-la
64 Mody Road, Tsimshatsui East
Tel: 721-2111
Fax: 723-8686

Nikko HK
72 Mody Road, Tsimshatsui East
Tel: 739-1111
Fax: 311-3122

Peninsula
Salisbury Rd., Tsimshatsui
Tel: 366-6251
Fax: 722-4170

Ramada Renaissance
8 Peking Rd., Tsimshatsui
Tel: 311-3311
Fax: 311-6611

The Regent
Salisbury Rd., Tsimshatsui
Tel: 721-1211
Fax: 739-4546

US$150-199 (HK$1170-1560)

Holiday Inn Golden Mile
50 Nathan Rd., Tsimshatsui
Tel: 369-3111
Fax: 369-8016

Holiday Inn Harbour View
70 Mody Road, Tsimshatsui East
Tel: 721-5161
Fax: 369-5672

Hyatt Regency
67 Nathan Rd., Tsimshatsui
Tel: 311-1234
Fax: 739-8701

Marco Polo
Harbour City, Tsimshatsui
Tel: 736-0888
Fax: 736-0022

Miramar
130 Nathan Rd., Tsimshatsui
Tel: 368-1111
Fax: 369-1788

New World
22 Salisbury Rd., Tsimshatsui
Tel: 369-4111
Fax: 369-9387

Omni Hong Kong
Harbour City, Canton Rd., Kowloon
Tel: 736-0888
Fax: 736-0011

Omni Marco Polo Hong Kong
Harbour City, Canton Rd.,
Tsimshatsui
Tel: 736-0888
Fax: 736-0022

Omni Prince
Harbour City, Tsimshatsui
Tel: 736-1888
Fax: 736-0066

Regal Airport
30-38 Sa Po Rd., Kowloon City
Tel: 718-0333
Fax: 718-4111

Regal Meridien
71 Mody Road, Tsimshatsui East
Tel: 722-1818
Fax: 723-6413

Royal Garden
69 Mody Rd., Tsimshatsui
Tel: 721-5215
Fax: 369-9976

Sheraton Hong Kong Hotel & Towers
20 Nathan Rd., Tsimshatsui
Tel: 369-1111
Fax: 739-8707

US$100-149 (HK$780-1170)

Ambassador
26 Nathan Rd., Tsimshatsui
Tel: 366-6321
Fax: 369-0663

Empress
17-19 Chatham Rd., Tsimshatsui
Tel: 366-0211
Fax: 721-8168

Fortuna
351-361 Nathan Rd., Yamati
Tel: 385-1011
Fax: 780-0011

Grand
14 Carnarvon Rd., Tsimshatsui
Tel: 366-9331
Fax: 723-7840

Grand Tower
627-641 Nathan Rd., Mongkok
Tel: 789-0011
Fax: 789-0945

Guangdong
18 Prat Ave., Tsimshatsui
Tel: 739-3311
Fax: 721-1137

Imperial
30-34 Nathan Rd., Tsimshatsui
Tel: 366-2201
Fax: 311-2360

International
33 Cameron Rd., Tsimshatsui
Tel: 366-3381
Fax: 369-5381

US$50-99 (HK$390-780)

Bangkok Royal
2-12 Pikem St., Jordan
Tel: 735-9181
Fax: 730-2209

King's
473-473A Nathan Rd., Yamati
Tel: 780-1281

Shamrock
223 Nathan Rd., Yamati
Tel: 735-2271
Fax: 736-7354

YMCA The Salisbury
Salisbury Rd., Tsimshatsui
Tel: 369-2211
Fax: 739-9315

NEW TERRITORIES

US$100-149 (HK$780-1170)

Regal Riverside
Tai Chung Kiu Rd., Shatin
Tel: 649-7878
Fax: 637-4748

Royal Park
8 Pak Hok Ting St., Shatin
Tel: 601-2111
Fax: 601-3666

Serviced apartments

Visitors staying in Hong Kong for an extended period should consider one-bedroom serviced apartments situated in apartment complexes on both Hong Kong Island and Kowloon and cost 40-50% less than a hotel. The HKTA can provide the names of some establishments (Tel: 801-7177, Fax: 801-4877).

LOCAL INFORMATION

Climate

Hong Kong's climate is sub-tropical with four distinct seasons during the year:

Spring: March through May, temperatures 14°-22°C, humidity often 80% and rain frequently.

Summer:	June through September, temperatures 30°-32°C, humidity often exceeds 90%.
Autumn:	October through December, temperatures average 25°C, humidity falls to 70%.
Winter:	December through February, temperatures 15°-21°C, humidity around 84% with fog and rain.

Typhoons affect Hong Kong between May and October almost annually but the early warning systems are very efficient.

Clothing

For business purposes, men wear suits or a jacket and tie all year round because most offices are air conditioned. Dress for businesswomen is also formal; trousers are rarely worn. Air conditioning in restaurants can sometimes be so fierce as to make wearing a jacket advisable.

Business hours

Most large businesses and government offices are open from 9 am to 5 pm, on weekdays with a lunch break between 1 pm and 2 pm. On Saturdays, the hours are 9 am to 1 pm. Small shops frequently open at 10 am and close as late as 10-11 pm. Most banks start work at 9 am but closing time varies between 3 pm and 5 pm,

depending on the bank. Auto-teller machines for banks and credit cards are widely available.

Holidays

Listed below are the main holidays. Those preceded by an asterisk follow the lunar calendar and fall on different dates each year.

January 1   New Year's Day.

*__Mid-February__   Chinese (Lunar) New Year.
(The first, second and third day of the Chinese New Year are public holidays.)

*__End of March__   Easter.
(A three-day public holiday beginning on Good Friday and ending on Easter Sunday.)

*__Early April__   Ching Ming Festival.
(A time to visit the graves of one's ancestors. Many people take a 3-4 day holiday; public transport is extremely crowded, especially at the border crossing into China.)

*__Early June__   Tuen Ng (Dragon Boat) Festival.

__June 15__   Queen's Birthday.
(Normally held on a Saturday in June. The Monday after is also a public holiday.)

__Last Monday in August__   Liberation Day.
(Commemorates the liberation of Hong Kong from Japan in WWII. The preceding Saturday is also a public holiday.)

*Late September Mid-Autumn (Moon) Festival.
(Because the festival begins at night, the following day is a public holiday.)

*Mid to late October Cheung Yeung Festival.

December 25, 26 Christmas and Boxing Day.

Business centers

In prime areas, including Central and Wanchai on Hong Kong Island and Tsimshatsui on Kowloon, there are over 20 business centers offering temporary office accommodation. Depending on room size, monthly rental charges range from HK$5,000-16,000. Offices can be rented by the hour for as little as HK$80 per hour. To find the closest center to your hotel, consult the "Business Services" section of the Yellow Pages, or the classified ads of the *South China Morning Post* and the *Hong Kong Standard*.

Hotel business centers cost a little less. Full secretarial services are offered at rates varying from HK$500-1,200 per day to HK$60-200 per hour. Other services provided by some hotels are portable telephone rentals (HK$20-300 per day), computer rentals (HK$120-1000 per day), conference room rentals (HK$50-200 per hour, HK$400-1400 per day), and private office rentals (HK$300-900 per day). Because many hotel business centers are closed on Sundays, it is best to check business center availability with the hotel of your choice before arriving in Hong Kong.

Electricity

The standard is 220V, 50Hz AC. Appliances designed for 110V will need a transformer. Plug adapters may also be necessary because Hong Kong's electrical outlets are designed to accommodate three round prongs. Inexpensive plug adapters are available in many appliance stores.

Weights and measures

Hong Kong's official system is the metric system. Local food markets, however, often use Chinese units such as the leung (37.5 grams) or the catty (600 grams). There are sixteen leungs to the catty.

Postal Service

The postal system is very efficient. All post offices are open Monday to Saturday from 8 am to 6 pm, and are closed on Sunday and public holidays. Allow five days for delivery of letters to Europe and North America.

Crime

Hong Kong is generally safe for foreigners. Pickpocketing is the main crime visitors must worry about. Tourist districts, like Tsimshatsui, are heavily patrolled by the police; those with red shoulder tabs can speak English.

Identification

Residents are required by law to carry HK Identification Cards at all times. Legally, visitors are also required to have some form of identification on them because the police are entitled to request proof of identity at any time. If you do not wish to carry around your passport, any other document bearing your picture (e.g., a driver's license) is acceptable.

ENTERTAINMENT

Restaurants

Hong Kong has an enormous number of high quality restaurants specializing not only in Chinese food but also in Thai, Indian, Vietnamese, Japanese and Western cuisine. One can also find fast food giants, such as McDonald's, Pizza Hut and Kentucky Fried Chicken. Smaller restaurants of uncertain quality are probably best avoided since they may prove too demanding on Western digestive systems. The Hong Kong Tourist Association (HKTA) publication, *Dining and Nightlife*, is one of the best sources of information on the subject as is their booklet, *The Official Hong Kong Guide,* re-issued every month. Both should be available at your hotel's front desk.

Nightlife

Hong Kong has innumerable drinking spots (with or without music), nightclubs, discos, *karaoke* bars, and girlie bars. These exist in every category, from the most sleazy (not necessarily the cheapest) to the most exclusive and lavish. Many of the up-market clubs are designed for groups or expense-account executives who seek a cosy atmosphere in which to relax in the company of hostesses.

Other entertainment

Both the *South China Morning Post* and the *Hong Kong Standard* contain detailed listings of daily events.

Shopping

Although Hong Kong is no longer the mecca for cheap goods it once was, the sheer variety of products available makes it one of Asia's best shopping centers. If you are looking for bargains, limit yourself to products manufactured in Hong Kong or in the PRC. HKTA has many useful booklets on this subject (e.g., *Shopping*, *Shopping Guide to Video Products* and *Shopping Guide to Jewelry*). HKTA publications are often available free of charge from the front desk of your hotel. Be warned that prices at duty-free shops at the airport are sometimes higher than in Hong Kong.

Markets and small shops

Haggling is possible here for those who pay cash. Reductions of 25% or more are possible but it is best that you first investigate the price of an item in the fixed-price department stores before trying to bargain with a small shop owner.

Sightseeing tours

Sightseeing tours are best handled through hotel reception desks. HKTA has many pamphlets given away at most hotels listing dozens of tours with brief descriptions, including routes, prices, times and whether meals are included. One tour we particularly recommend for exporters of consumer products wishing to better understand the living conditions of most Hong Kong consumers is "Housing Tour and Home Visits". HKTA can be reached at Tel. 801-7177.

HEALTH CARE

Hospitals

All major hotels have clinics and resident nurses. Hong Kong has no national health insurance and all medical treatment must be paid by the patient in advance. Public hospitals charge low fees but Hong Kong residents pay less than foreign visitors.

Public hospitals with 24-hour casualty service include:

Queen Elizabeth Hospital
Wylie Rd., Yamati, Kowloon
Tel: 710-2111

Queen Mary Hospital
Pokfulam Rd., Hong Kong Island
Tel: 817-9463

Private hospitals can also be excellent but charge more because they must make a profit. Some of the better ones include:

Adventist
40 Stubbs Rd., Wanchai
Hong Kong Island
Tel: 574-6211

Baptist
222 Waterloo Rd., Kowloon Tong
Tel: 337-4141

Dentists

Dentists are too few in number for Hong Kong's population and their fees are consequently very high. It is best to have your teeth examined before going to Hong Kong.

Pharmacies

Pharmacies in Hong Kong are usually open from 9 am to 6 pm, with some closing only at 8 pm. Watson stores are all over Hong Kong. 7-Eleven stores often stock basic pharmaceuticals and are open 24 hours.

DEPARTURE

Airport tax

Hong Kong's airport tax is HK$150.

Checked in luggage

To make sure you don't miss your flight because of traffic jams, it is best to arrive at Kai Tak Airport two hours before departure. Please note too that announcements are only made inside the departure area. If you must leave it for any reason, keep an eye on the information monitors and allow sufficient time to complete immigration, customs and boarding formalities.

ANNEXES

ANNEX 1

CENSUS AND STATISTICS DEPARTMENT (HK) PUBLICATIONS

The following is a list of publications and prices (in HK$), the only one available to us in June 1992, dated December 1991. More up-to-date lists can be obtained at the following addresses:

WHERE TO OBTAIN PUBLICATIONS

Government Publications Centre
General Post Office Building
Ground Floor, Connaught Place
Central, Hong Kong

Census & Statistics Department
Wanchai Tower 1, 19/F
12 Harbour Road
Wanchai, Hong Kong

Orders by post/Subscriptions for periodicals
Send mail orders and annual subscriptions for periodicals to:

Director of Information Services
1 Battery Path, Ground Floor
Central, Hong Kong

(Please quote full title and enclose cheque or remittance payable to cover cost and postage to Hong Kong government.)

PUBLICATIONS

General Digests

Hong Kong Monthly Digest of Statistics (monthly, $48)

Collection of main monthly/quarterly data series from all government departments.

Hong Kong Annual Digest of Statistics (annually, $98)
The most comprehensive collection of data series from government departments and data from major censuses and surveys.

Hong Kong Social and Economic Trends (bi-annually, $40)
Selection of tables and charts on major social and economic statistics.

Hong Kong in Figures (annual, free, upon request only)
Collection of the most important summary statistics for describing Hong Kong.

General Economic

Hong Kong Economic Trends (monthly, $1.50)
Summary statistics of major economic indicators on trade, finance, prices and retailing, construction, production, transport, tourism and labor.

Report on Quarterly Business Survey (quarterly, $8)
   Summary of opinions of business undertakings concerning expectations for the following quarter.

Monthly Survey of Employment, Payroll and Orders-on-hand (monthly, $4.50)
   Short-term economic indicators on employment, payroll and orders-on-hand from a monthly panel survey.

External trade

Hong Kong Trade Statistics (monthly, $100 per volume)
   Figures are published in two volumes, each for imports and exports. Monthly figures for individual commodity items are analyzed by country whereas the figures for annual supplements are analyzed country by commodity.
   [This publication is also available in the computer microfiche (COM) in three separate series for imports (monthly $25, annual supplement $60), domestic exports (monthly $25, annual supplement $65) and re-exports (monthly $35, annual supplement $90).]

Hong Kong External Trade (monthly, $33.50)
   Summary tables on external trade performance, and important changes in the direction and content of trade, trade statistics analyzed by principal products, main markets, and main markets by principal products, as well as statistics on airborne and seaborne trade, and quarterly statistics on trade with China of an outward processing nature.

Hong Kong Trade Index Numbers (monthly, $4)
   Trade index numbers showing external trade performance measured in terms of value, quantity and unit value.

Hong Kong Shipping Statistics (annual, $50; quarterly, $34.50)
   Shipping statistics (in terms of number and capacity) and seaborne cargo throughput statistics (in terms of gross weight and volume) analyzed by country/port, shop characteristics, principal commodity, etc. Supplementary statistics on cargo throughput in terms of gross weight by other modes of transport and container throughput of leading container ports are also shown.

National Income

Quarterly Estimates of Gross Domestic Product (quarterly, $2)
   Latest quarterly estimates of Gross Domestic Product and its expenditure components.

Estimates of Gross Domestic Product (annual, $61)
   Estimates of the Gross Domestic Product of Hong Kong using the expenditure approach and the production approach with a full description of the compilation methods and data sources used in arriving at the estimates.

Gross Domestic Product: Quarterly Estimates and Revised Annual Estimates (published in August 1991) (ad hoc, $12)

Quarterly estimates (from the 1st quarter of 1973 to the 1st quarter of 1991) and revised annual estimates (from 1966 to 1990) of the Gross Domestic Product of Hong Kong with a full description of the concepts, data sources, estimation methods and the features of the quarterly estimates as well as the scope and background of the revision of the annual GDP estimates.

Labor Force, Employment, Wages

General Household Survey Labour Force Characteristics (quarterly, $13.50)

Analysis of labor force participation, unemployment and underemployment, with tables showing the characteristics of the labor force.

Quarterly Report of Employment, Vacancies and Payroll Statistics (quarterly, $13.50)

Summary statistics and analysis on establishments, persons engaged, vacancies and monthly payroll classified by industry groups/sectors.

Employment and Vacancies Statistics (Detailed Tables, annual)

Detailed employment and vacancies statistics analyzed by census district and employment size of establishments in three separate

volumes for the industrial sector ($71), the services sector ($152), and the distribution sector ($54).

Report on Half-Yearly Survey of Wages, Salaries and Employee Benefits (half-yearly, in 2 volumes, $9.50–$21 per volume)

Nominal and real wage indexes for manual and non-manual workers analyzed by occupational groups and industries/sectors. Average daily wage and average monthly salary for common occupations in various industries.

Report of Salaries and Employee Benefits Statistics – Managerial and Professional Employees (Excluding Top Management) (annual, $25)

Nominal and real salary indexes for managerial and professional employees analyzed by major economic sectors and by occupational groups. Average monthly salary for common occupations in various economic sectors and statistics on entitlement to fringe benefits.

Quarterly Report of Employment and Vacancies at Construction Sites (quarterly, $8.50)

Statistics on number of construction sites, manual workers engaged and vacancies of manual workers at building and construction sites in both private and public sectors.

Consumer Prices and Household Expenditure

Consumer Price Index Report (annual, $27, monthly, $7)

Statistics and analysis of price movement of consumer goods and services.

1989/90 Household Expenditure Survey and the Rebasing of the Consumer Price Indexes (ad hoc, English version $56, Chinese version $60)
Analysis of household expenditure data, and description of the concepts of the Consumer Price Indexes and salient features of the 1989/90-based Consumer Price Index series.

Industrial Production

Survey of Industrial Production (annual, $26)
Information on the structure and operating characteristics of industrial establishments.

Report on the Quarterly Index of Industrial Production (quarterly, $3)
Quarterly indexes to show the up-to-date trends of net output in various manufacturing industries.

Report on Textile Production Statistics (quarterly, $2)
Figures on textile production, materials consumed, stock, and index of textile production in the spinning and weaving industries.

Distribution and Services

Survey of Wholesale, Retail and Import/Export Trades, Restaurants and Hotels (annual, $24)
Information on the structure and operating characteristics of establishments in these trades.

Report on Quarterly Survey of Restaurant Receipts and Purchases (quarterly, $1.50)
Figures indicating short-term changes in the value and volume of receipts of various types of restaurants.

Report of Transport & Related Services (annual, $22)
Information on the structure and operating characteristics of establishments in the transport industry.

Survey of Storage, Communications, Financing, Insurance and Business Services (annual, $25)
Information on the structure and operating characteristics of establishment in these trades.

Building and Construction

Survey of Building, Construction and Real Estate Sectors (annual, $26)
Information on the structure and operating characteristics of all building and civil engineering contractors, architectural, surveying and project engineering firms, and real estate developers.

Report on the Quarterly Survey of Construction Output (quarterly, $3)
Quarterly data on the value of construction work performed by building and civil engineering contractors.

Energy

Hong Kong Energy Statistics (annual, $23/quarterly, $6.50)

Statistics on supplies and consumption of oil products, coal products, electricity and gas in terms of volume, prices sources and storage capacity.

Population

Hong Kong 1991 Population Census
  First series of publications:
    Summary Results
    –contains 35 tables, 3 charts and commentaries summarizing the findings of the 1991 Population Census ($20)
    Tabulation for District Board Districts and Constituency Areas
    –Population by Age and Sex ($31)
    –Living Quarters, Households and Population by Type of Living Quarters ($24)
    Tabulations for Tertiary Planning Units
    –Population by Age and Sex ($22)
    –Living Quarters, Households and Population by Type of Living Quarters ($22)
    Maps
    –Boundary Maps Complementary to Tabulations for District Board Districts and Constituency Areas ($48)
    –Boundary Maps Complementary to Tabulations for Tertiary Planning Units ($72)

Social

Special Topics Report – Social Data Collected by the General Household Survey (Report I $15; Report II $16.50; Report III $12; Report IV $13; Report V $29; Report VI $28; Report VII $46)
  Results of the enquiries on a variety of social topics in the General Household Survey. Examples are part-time employment, attending cultural performances, part-time education, country parks, heritage preservation, expenditure on public transportation, taxi waiting time, domestic helpers, Hong Kong residents working in China, employed persons actively seeking other employment, sick leave pattern and maternity leave pattern of employees, hospitalization, doctor consultation, cigarette smoking pattern, toy safety, family life education service and awareness of Family Service Centres.

Crime and Its Victims in Hong Kong 1989 (ad hoc, $54)
  Figures from the 1990 Crime Victimization Survey showing the number and characteristics of crimes and its victims in 1989.

ANNEX 2

HKTDC'S INTERNATIONAL NETWORK

Hong Kong

Head Office	38th Fl., Office Tower, Convention Plaza 1 Harbour Road, Hong Kong Tel: (852) 584-4333 Fax: (852) 824-0249 Cable: CONOTRAD HONG KONG Telex: 73595 CONHK HX
Advertising Dept.	Room 1303, Block A, 13/F. Sea View Estate, 2 Watson Rd. North Point, Hong Kong Tel: (852) 566-7292 Fax: (852) 887-6490
TDC Datashops	G/F, Trade Department Tower 700 Nathan Road Mongkok, Kowloon, Hong Kong Tel: (852) 390-0276
GPO	G/F, 2 Connaught Pl. Hong Kong Tel: (852) 877-1786
Fou Wah Centre	1st Fl., Shop 4A, 210 Castle Peak Rd, Tsuen Wan, New Territories, Hong Kong Tel: (852) 498-1602 Fax: (852) 413-1942
Chung Nam Centre	Ground Floor, 414 Kwun Tong Road Kwun Tong, Kowloon, Hong Kong Tel: (852) 341-2314/7 Fax: (852) 343-2545

North America

(For trade enquiries in the US, call 1-800-TDC-HKTE)

Chicago	333 N. Michigan Ave., Suite 2028 Chicago, IL, 60601 Tel: (312) 726-4515 Fax: (312) 726-2441
Dallas Postal Address:	Suite 120, World Trade Center 2050 Stemmons Freeway Dallax, TX 75207 PO Box 58329, Dallas, TX 75258 Tel: (214) 748-8162 Fax: (214) 742-6701
Los Angeles	Los Angeles World Trade Center 350 S. Figueroa St., Suite 282 Los Angeles, CA 90071-1386 Tel: (213) 622-3194 Fax: (213) 613-1490
Miami	Courvoisier Center, Suite 402 501 Brickell Key Drive Miami, FL 33131 Tel: (305) 577-0414 Fax: (305) 372-9142
New York	219 East 46th Street, New York, NY 10017 Tel: (212) 838-8688 Fax: (212) 838-8941
San Francisco	c/o Hong Kong Economic & Trade Office 222 Kearny Street, 4th Floor, Suite 402 San Francisco, CA 94108 Tel: (415) 677-9038 Fax: (415) 421-0646
Toronto	Suite 1100, National Building 347 Bay Street Toronto, Ontario M5H 2R7 Tel:(416) 366-3594 Fax: (416) 366-1569

Vancouver

Suite 700, 1550 Alberni Street
Vancouver, British Columbia
V6G 1A3
Tel: (604) 685-0883
Fax: (604) 669-3784

Central America

Mexico City

Manuel E. Izaguirre #13,
3er piso, Ciudad Satelite,
Mexico City 53310, Mexico
Tel: 52-5-572-41-13, 572-41-31
Fax: 52-5-393-59-40

Panama City

Condominio Plaza Internacional, Primer Alto
Oficina No. 27, Edificio del Banco Nacional de
Panama
Via Esplana y Calle 55 Panama City, Republica
de Panama
Tel: 507-69-5894, 69-5611, 69-5109
Fax: 507-69-6183

Asia

Bangkok

20/F Pacific Tower
21 Vibhavadi Rangsit Road
Bangkok 10900, Thailand
Tel: 66-2-273-8800
Fax: 66-2-273-8880

Beijing

Room 901, 9th Floor, CITIC Building
19, Jianguomenwai Dajie
Beijing, China
Tel: 86-01-500-2255 ext. 3910, 512-8661
Fax: 86-01-500-3285

Nagoya

Sakae-Machi Bldg., 4/Fl.
3-23-31 Nishiki, Naka-ku
Nagoya, 460 Japan
Tel: 81-052-971-3626
Fax: 81-052-962-0613

Osaka	Osaka Ekimae Dai-San Bldg. 6th Floor, 1-1-3 Umeda, Kita-ku Osaka 530 Japan Tel: 81-06-344-5211 Fax: 81-06-347-0791
Seoul	720-721 KFSB Building 16-2, Yoidodong, Youngdeungpoku Seoul, Korea Tel: 82-02-782-6115/7 Fax: 82-02-782-6118
Shanghai	Room 1004, 10th Fl. Shanghai Union Building 100 Yanan Dong Lu Shanghai 200002, China Tel: 86-21-326-4196,326-5935 Fax: 86-21-328-7478
Singapore	20 Kallang Avenue, 2nd Fl. Pico Creative Centre Singapore 1233 Tel: 65-293-7977 Fax: 65-292-7767
Taipei	315 Sung Chiang Road, 7th Fl. Taipei, Taiwan Tel: 886-02-516-6085 Fax: 886-02-502-2115
Tokyo	Toho Twin Tower Bldg, 4th Fl., 1-5-2 Yurakucho Chiyoda-ku, Tokyo 100 Tel: 81-03-3502-3251/5 Fax: 81-03-3591-6484

Australia

Sydney	71 York Street Sydney, NSW, 2000 Australia Tel: 61-02-299-8343 Fax: 61-02-290-1889

Europe

Amsterdam

Prinsengracht 771, G/F
1017 JZ Amsterdam
The Netherlands
Tel: 31-020-627-7101
Fax: 31-020-622-8529

Athens

2, Vassileos Alexandrou Str.
Athens 11634, Greece
Tel: 30-1-724-0594
Fax: 30-1-724-8922

Barcelona

Balmes 184
Barcelona 08006 Spain
Tel: 34-3-415-8382,34-3-415-6628
Fax: 34-3-416-0148

Budapest

H-117 Budapest, Kaposvar u.
5-7 Hungary
Tel: 36-1-181-3398
Fax: 36-1-161-2458

Frankfurt

Kreuzerhohl 5-7, D-6000
Frankfurt/Main 50, Germany
Tel: 49-069-586-011
Fax: 49-069-589-0752

Istanbul

Piyalepasa Bulvari, Kastel Is Merkesi
D. Blok Kat: 5,80370
Piyalepasa, Istanbul, Turkey
Tel: 90-1-154-98-67
Fax: 90-1-154-98-67

London

Swire House, Ground Floor
59 Buckingham Gate
London, SWIE 6AJ, England
Tel: 44-071-828-1661
Fax: 44-071-828-9976
For trade enquiries in the UK, call 0800-282-980

Milan

2 Piazetta Pattari
20122 Milan, Italy
Tel: 39-02-865-405
Fax: 39-02-860-304

Paris	18, rue d'Aguesseau
	75008 Paris, France
	Tel: 33-01-47-42-41-50
	Fax: 33-01-47-42-77-44
Stockholm	Kungsgatan 6,
	S-111 43 Stockholm, Sweden
	Tel: 46-08-100-677
	Fax: 46-08-723-1630
Vienna	Rotenturmstrasse 1-3/8/24 A-1010
	Vienna, Austria
	Tel: 43-01-533-98-18
	Fax: 43-01-535-31-56
Zurich	Seestrasse 135
	Zurich, Switzerland
	Tel: 41-01-281-31-91
	Fax: 41-01-281-31-91

ANNEX 3

SOME TRANSNATIONAL DISTRIBUTORS

We have limited ourselves only to those specialties a given distributor may have in Hong Kong and the PRC who may very well carry different product lines elsewhere in Southeast Asia or in Europe.

SOME LARGER ONES

THE EAST ASIATIC COMPANY (EAC)

The East Asiatic Company (Hong Kong) Ltd.
11/F, Great Eagle Centre
23 Harbour Road, Wanchai, Hong Kong
Tel: (852) 586-6888
Fax: (852) 827-5180

Specialties: Graphics, transportation, commodity trading, consumer products (Hong Kong and China).

Comment: Founded in 1897 in Copenhagen, Denmark, EAC now has 135 companies in more than 50 countries.

EDWARD KELLER LTD.

Edward Keller Ltd.
36/F, Windsor House
311 Gloucester Road
Causeway Bay, Hong Kong
Tel: (852) 895-0888
Fax: (852) 577-1057

Specialties: Pharmaceuticals, soft goods, consumer products, industrial equipment.

Comment: Founded in the late 1860s, Edward Keller is a Swiss company best known in Hong Kong for its distribution of fast-moving consumer products.

THE GETZ GROUP

Getz Bros. & Co., Inc.
12/F, Baskerville House
22 Ice House Street
Central, Hong Kong
Tel: (852) 869-0585
Fax: (852) 524-9670

Specialties: Consumer products, building materials, office furniture, light industry machinery, chemicals, plastic molds.

Comment: Getz is based in San Francisco and provides international marketing and distribution services in Western and Central Europe, Asia, Australia and the Indian subcontinent.

THE HAGEMEYER GROUP

Hagemeyer (Hong Kong) Ltd.
9/F, Warwick House, West Wing
28 Tong Chong Street
Quarry Bay, Hong Kong
Tel: (852) 563-5101
Fax: (852) 565-8721

Specialties: Household products, electrical appliances, hotel and restaurant furniture and equipment, cosmetics, wine and spirits.

Comment: Headquartered in the Netherlands, Hagemeyer's involvement in China is limited to the foreign currency market. It was founded in 1900.

THE JARDINE GROUP

Jardine Pacific Limited
28/F, World Trade Centre
Causeway Bay, Hong Kong
Tel: (852) 837-3888
Fax: (852) 895-6909

Specialties: Consumer products, specialty retailing (IKEA, Optical Shops), system products, industrial and agricultural products, chain restaurants (Pizza Hut, Ziggler and Taco Bell), and motor vehicles (Mercedez-Benz).

Comment: Founded in 1832, Jardine's is Hong Kong's most famous *hong*. Jardine Pacific acts as the trading and services arm of Jardine Matheson Holdings Ltd. The group's buying power is enhanced by the fact that it operates, through a separate subsidiary, Dairy Farm, a supermarket chain (Wellcome), a drugstore chain (Mannings) and a chain of highly successful convenience stores (7-Eleven). It is also involved in the PRC market.

THE JDH GROUP (INCHCAPE)

JDH Centre
2 On Ping Street
Siu Lek Yuen, Shatin
New Territories, Hong Kong
Tel: (852) 635-5600
Fax: (852) 637-5642

Specialties: Consumer products (durable and fast-moving), medical equipment, pharmaceuticals, business machines, wine and spirits, industrial equipment, and automobiles (Toyota).

Comment: Headquartered in London, JDH is a massive and highly diversified group of companies acting in more than 60 countries. It employs over 49,000 people of which 14,000 live in the Asia Pacific region. It is very involved in the PRC market.

THE JEBSEN GROUP

Jebsen & Co. Ltd.
2-38, Yun Ping Road
28/F, Caroline Centre
Causeway Bay, Hong Kong
Tel: (852) 822-3877
Fax: (852) 882-1399

Specialties: Power engineering and machinery, printing and packaging machinery, medical and scientific equipment, telecommunications equipment, consumer durables and automobiles. Its technical division is heavily involved in the PRC.

Comment: Founded in 1895, Jebsen is one of the oldest trading groups operating in Asia. Its management is mainly Danish.

THE SCHMIDT GROUP

Schmidt & Co., (HK) Ltd.
18/F Great Eagle Centre
23 Harbour Road, Wanchai, Hong Kong
Tel: (852) 507-0222
Fax: (852) 827-5656

Specialties: Medical, scientific, production, engineering, telecom, test and measurement, and photographic equipment; electronic components and CAD/CAM systems.

Comment: It has five service centers in China.

THE SIME DARBY GROUP

Sime Darby Hong Kong Ltd.
16/F, East Wing, Hennessy Centre
500 Hennessy Road, Causeway Bay, Hong Kong
Tel: (852) 895-0777
Fax: (852) 890-5896

Specialties: Motor vehicles (Mitsubishi, BMW, Ford, Suzuki), construction equipment, and general trading.

Comment: Originally a UK company, Sime Darby is now entirely owned by Asian (mainly Malaysian) interests. Singapore is the main hub of its marketing and distribution activities.

THE SWIRE GROUP

Swire Pacific Ltd.
Swire House
9 Connaught Road, Central, Hong Kong
Tel: (852) 840-8097
Fax: (852) 810-6563

Specialties: Insurance, property, trading, aviation (it controls Cathay Pacific), shipping, distribution, and manufacturing.

Comment: A large UK-owned conglomerate. Two of its subsidiaries are particularly involved in distribution: Swire Systems Ltd. and Swire Engineering Ltd.

SOME SMALLER ONES

ARNHOLD & CO. LTD.

Arnhold & Co. Ltd.
Victoria Centre 6/F
15 Watson Road, Hong Kong
Tel: (852) 807-9400
Fax: (852) 806-1744

Specialties: Plumbing and bathroom products.

CLARIDGE HOUSE LTD.

Claridge House Ltd.
2/F Lok Moon Commercial Centre
29 Queen's Road East, Hong Kong
Tel: (852) 527-8121-4
Fax: (852) 529-1564

Specialties: Playground and park equipment, gymnasium and sports equipment.

GOODMAN MEDICAL SUP-PLIES LTD.

Goodman Medical Supplies Ltd.
335 Nathan Road, 8/F
Kowloon, Hong Kong
Tel: (852) 332-4455
Fax: (852) 710-9696

Specialties: Medical equipment.

JACOBSON VAN DEN BERG

Jacobson van den Berg (Hong Kong) Ltd.
237 Lockhart Road
Wanchai, Hong Kong
Tel: (852) 828-9328
Fax: (852) 828-9388

Specialties: Pharmaceuticals, plastics, machinery.

THE LAMKO GROUP

Lamko Tool & Mold Consulting (H.K.) Co., Ltd.
1/F, Ngai Wong Commercial Building
11-13 Mongkok Road, Mongkok
Kowloon, Hong Kong
Tel: (852) 398-1262
Fax: (852) 398-1261

Specialties: Distribution of computer numeric-controlled machine tools and associated products, engineering equipment, electrical appliances, and some consumer products.

WIGGINS TEAPE (HONG KONG) LTD.

Wiggins Teape (Hong Kong) Ltd.
7/F Kingsford Industrial Centre, Unit A
6 Cho Yuen Street, Yau Tong
Kowloon, Hong Kong
Tel: (852) 775-0196
Fax: (852) 775-7132

Specialties: Paper products.

ANNEX 4

TOP CONSULTING AND MARKET RESEARCH FIRMS

Some of these firms are limited to market research; others are primarily interested in strategy consulting.

Asian Market Intelligence Ltd. (AMI)
28/F Shui On Centre
8 Harbour Road
Wanchai, Hong Kong
Tel: (852) 824-2832
Fax: (852) 824-2853

Asian Commercial Research Ltd. (ACR)
3c, 68 Johnston Road
Trust Tower, Wanchai, Hong Kong
Tel: (852) 527-6821
Fax: (852) 865-7390

The Boston Consulting Group International Inc.
38/F, West Tower
Bond Centre, Hong Kong
Tel: (852) 868-3161
Fax: (852) 524-1517

Business International Asia-Pacific Ltd.
10/F, Luk Kwok Centre
72 Gloucester Road, Hong Kong
Tel: (852) 529-0833
Fax: (852) 865-1554

Frank Small & Associates Limited
19/F, OTB Building
160 Gloucester Road, Wanchai

Hong Kong
Tel: (852) 891-1175
Fax: (852) 834-2261

International Research Associates (HK) Ltd. (INRA)
Rm 704-5, C.C. Wu Building
302 Hennessy Road, Hong Kong
Tel: (852) 893-3421
Fax: (852) 834-5102

Marketing Decision Research (Pacific) Ltd. (MDR)
12B CDW Building, 388 Castle Peak Road
Tsuen Wan, New Territories
Hong Kong
Tel: (852) 413-5777
Fax: (852) 414-4239

MBL Asia-Pacific Ltd.
Room 1606, Eastern Centre
1065 King's Road, Hong Kong
Tel: (852) 811-9668
Fax: (852) 811-9988

McKinsey & Company
18/F, Two Exchange Square
8 Connaught Place, Central
Hong Kong
Tel: (852) 868-1188
Fax: (852) 845-9985

SRG China Consultancy
7/F Warwick House, East Wing
28 Tong Chong Street
Quarry Bay, Hong Kong
Tel: (852) 880-3388
Fax: (852) 565-0418
          565-9560

SRH
Survey Research Hong Kong Ltd.
7/F Warwick House, East Wing
28 Tong Chong Street
Quarry Bay, Hong Kong
Tel: (852) 880-3388
Fax: (852) 565-0418

The source of much of the statistical data contained in "The Hong Kong Consumer" and "The PRC Customer: Consumer Products".

Pac Rim
8/f, California Tower
30-32 D'Aquilar Street
Central, Hong Kong
Tel: (852) 526-4061
Fax: (852) 810-4845

**Technomic Consultants
International**
14/F China Underwriters Centre
88 Gloucester Road, Hong Kong
Tel: (852) 511-2831
Fax: (852) 519-9503

ANNEX 5

TOP LAW FIRMS

Baker & McKenzie
14/F Hutchison House
10 Harcourt Road, Hong Kong
Tel: (852) 846-1888
Fax: (852) 845-0476

Boughton, Peterson, Yang, Anderson
12/F Prince's Building
10 Chater Road, C., Hong Kong
Tel: (852) 877-3088
Fax: (852) 525-1099

Clifford Chance
30/F Jardine House
1 Connaught Place, Hong Kong
Tel: (852) 810-0229
Fax: (852) 810-4708

Clyde & Co.
19/F Tower Two
Admiralty Centre
Harcourt Road, Hong Kong
Tel: (852) 529-0017
Fax: (852) 865-4259

Deacons
3/F-6/F Alexandra House
Chater Road
Hong Kong
Tel: (852) 825-9275
Fax: (852) 845-9126

Johnson, Stokes & Master
17/F Prince's Building
10 Chater Road
Hong Kong
Tel: (852) 843-2211
Fax: (852) 845-1735, 845-9121

Slaughter & May
27/F, 2 Exchange Square
Central, Hong Kong
Tel: (852) 521-0551
Fax: (852) 845-9079, 845-2125

Stevenson, Wong & Co.
19/F, 9 Queen's Road Central
Hong Kong
Tel: (852) 526-6311
Fax: (852) 845-0638, 810-5868
Authors of "Elements of a
Distribution Agreement".

Stikeman, Elliott
Suite 1103, China Building
29 Queen's Road Central
Hong Kong
Tel: (852) 526-5531
Fax: (852) 845-9076
A Canadian-based global law
firm with offices in Montreal,
Toronto, Ottawa, Calgary,
Vancouver, Hong Kong, Taipei,
London and New York.

ANNEX 6

TOP ACCOUNTING FIRMS

Arthur Andersen
25/F Wing On Centre
111 Connaught Road
Central, Hong Kong
Tel: (852) 852-0222
Fax: (852) 875-0548

Bird Cameron
5/F On Lane Centre
On Lane St.
Hong Kong
Tel: (852) 526-2191
Fax: (852) 810-0502

Byrne & Co.
40/F Bond Centre East Tower
89 Queensway
Hong Kong
Tel: (852) 840-1188
Fax: (852) 840-0789

Charles Mar Fan & Co.
11/F Belgian House
77 Gloucester Rd.
Hong Kong
Tel: (852) 520-0333
Fax: (852) 529-4347

Coopers & Lybrand
23/F Sunning Plaza
10 Hysan Avenue
Hong Kong
Tel: (852) 839-4321
Fax: (852) 765-5356

Deloitte Ross Tohmatsu
26/F Wing On Centre
111 Connaught Rd. C.
Hong Kong

Tel: (852) 545-0303
Fax: (852) 541-1911

Ernst & Young
15/F Hutchison House
10 Harcourt Rd. C.
Hong Kong
Tel: (852) 846-9888
Fax: (852) 868-4432
Author of our chapter,
"Taxation in Hong Kong".

Glass, Radcliffe & Co.
806 Yu Yuet Lai Bldg.
43 Windham St.
Hong Kong
Tel: (852) 522-7259
Fax: (852) 810-1417

Hodgson Impey Chong
5/F The Chinese Club Bldg.
21 Connaught Rd. C.
Hong Kong
Tel: (852) 810-8333
Fax: (852) 810-1948

Horwath & Co.
606 Bank of America Tower
12 Harcourt Rd.
Hong Kong
Tel: (852) 526-2191
Fax: (852) 810-0502

KPMG Peat Marwick
8/F Prince's Building
10 Chater Road, Central
Hong Kong

Tel: (852) 522-6022
Fax: (852) 854-2588

Kwan, Wong, Tan & Fong
50/F Hopewell Centre
183 Queen's Rd. E.
Hong Kong
Tel: (852) 528-4111
Fax: (852) 865-6616

Nelson Wheeler
38/F Hopewell Centre
183 Queen's Rd. E.
Hong Kong
Tel: (852) 527-5123
Fax: (852) 529-6890

Pannell Kerr Forster
901 One Pacific Place
88 Queensway
Hong Kong
Tel: (852) 842-4000
Fax: (852) 868-1605

Price Waterhouse
22/F Prince's Building
5 Ice House
Hong Kong
Tel: (852) 826-2111
Fax: (852) 810-9888

Sanford Yung & Co.
23/F Sunning Plaza
Hysan Ave.
Hong Kong
Tel: (852) 839-4321
Fax: (852) 576-5356

Spicer & Oppenheim
2004 Hong Kong Club Bldg.
3A Chater Rd.
Hong Kong
Tel: (852) 521-0421
Fax: (852) 2734

ANNEX 7

TRADE JOURNALS

*Of the many trade journals we consulted while researching this book, four
stood out as particularly useful.*

Publication	Annual Subscription Rate	Publisher
Business China	US$ 620/24 issues	Business International 10/F Luk Kwok Centre 72 Gloucester Road Hong Kong Tel: (852) 529-0833 Fax: (852) 865-1554
The China Business Review (the source of our chapter, "The Birth of Greater China")	US$ 96/6 issues	The China Business Review 1818 N. St., NW Suite 500 Washington, DC 20036 USA Tel: (202) 429-0340 Fax: (202) 775-2476
The China Trade Report (the source of our section, "Avon's Finishing School")	US$ 375/12 issues	The China Trade Report GPO Box 160 Hong Kong Tel: (852) 832-8300 Fax: (852) 572-2436
Far Eastern Economic Review	US$ 149/51 issues	Far Eastern Economic Review GPO Box 160 Hong Kong Tel: (852) 832-8300 Fax: (852) 572-2436

ANNEX 8

CHAMBERS OF COMMERCE AND TRADE ASSOCIATIONS

CHAMBERS OF COMMERCE

American Chamber of Commerce (AmCham)
10/F, 1030 Swire House
7 Chater Rd.
Central, Hong Kong
Tel: (852) 526-0165
Fax: (852) 810-1289
Mail: GPO Box 355, Hong Kong

Australian Chamber of Commerce
701A, 7/F, Euro Trade Centre
13-14 Connaught Rd.
Central, Hong Kong
Tel: (852) 522-5054
Fax: (852) 877-0860

British Chamber of Commerce in Hong Kong
1712 Shui On Centre
8 Harbour Rd.
Central, Hong Kong
Tel: (852) 824-2211
Fax: (852) 824-1333

Canadian Chamber of Commerce
13/F, One Exchange Square
8 Connaught Place
Central, Hong Kong
Tel: (852) 526-3207
Fax: (852) 845-1654

Chinese General Chamber of Commerce
7/F, 24-25 Connaught Rd.
Central, Hong Kong
Tel: (852) 525-6385
Fax: (852) 845-2610
Tlx: 89854 CGCC HX

French Business Association of Hong Kong
18/F, Tower 2
Admiralty Centre
18 Harcourt Rd.
Central, Hong Kong
Tel: (852) 866-1007, 523-6818
Fax: (852) 861-0806
Tlx: 76715 UAPHK HX

Hong Kong General Chamber of Commerce
22/F, United Centre
95 Queensway, Hong Kong
Tel: (852) 529-9229
Fax: (852) 866-2035

Kowloon Chamber of Commerce
3/F, Kowloon Chamber of Commerce Bldg.
2 Liberty Avenue
Homantin, Kowloon
Tel: (852) 760-0393
Fax: (852) 761-0166

TRADE ASSOCIATIONS

Association of Accredited Advertising Agents of HK (4As)
504-505 Dominion Centre
43-59 Queen's Rd. East
Wanchai, Hong Kong
Tel: (852) 529-9656
Fax: (852) 861-0375

Canadian Standards Association
10 Dai Wang St.
Taipo Industrial Estate
Taipo, New Territories
Tel: (852) 664-2872
Fax: (852) 665-5033

Chinese Manufacturers' Association of Hong Kong
3-5/F, CMA Bldg.
64-66 Connaught Rd.
Central, Hong Kong
Tel: (852) 545-6166, 542-8600
Fax: (852) 541-4541
Tlx: 63526 MAFTS HX

Design Council of Hong Kong
Rm 407, Hankow Centre
5-15 Hankow Rd.
Tsimshatsui, Kowloon
Tel: (852) 723-0818
Fax: (852) 721-3494
Tlx: 30101 FHKI HX

Federation of Hong Kong Industries
Rm 408, Hankow Centre
5-15 Hankow Rd.
Tsimshatsui, Kowloon
Tel: (852) 723-0818
Fax: (852) 721-3494
Tlx: 30101 FHKI HX

Federation of Hong Kong Garment Manufacturers
25 Kimberley Rd., 4/F
Tsimshatsui, Kowloon
Tel: (852) 721-1383/6
Fax: (852) 311-1062

HK Advertisers' Association (2As)
Rm 1002
Cameron Commercial Centre
458-468 Hennessy Rd.
Causeway Bay, Hong Kong
Tel: (852) 832-9321
Fax: (852) 838-1595

HK Association of Freight Forwarding Agents
M/F, AHAFA Cargo Centre
12 Kai Shun Rd.
Kowloon Bay, Kowloon
Tel: (852) 796-3121
Fax: (852) 796-3719

HK Chinese Importers' & Exporters' Association
Champion Building, 7-8/F
287-291 Des Voeux Rd.
Central, Hong Kong
Tel: (852) 544-8474, 545-5998
Fax: (852) 544-4677

HK Coalition of Services Industries
c/o Hong Kong General Chamber of Commerce

HK Computer Society
14/F, Evernew House
485 Lockhart Rd.
Causeway Bay, Hong Kong
Tel: (852) 834-2228
Fax: (852) 834-3003

HK Construction Association
180-182 Hennessy Rd., 3/F
Wanchai, Hong Kong
Tel: (852) 572-4414
Fax: (852) 572-7104

HK Designers' Association
Rm 407-411, Hankow Centre
5-15 Hankow Rd.
Tsimshatsui, Kowloon
Tel: (852) 723-0818
Fax: (852) 721-3494

HK & M Contractors' Association
Quarry Bay Reclamation Area
Hoi Tai St.
Quarry Bay, Hong Kong
Tel: (852) 565-5411

Hong Kong Electronics Association
Rm 1806-8 Beverly House
93-107 Lockhart Rd.
Wanchai, Hong Kong
Tel: (852) 866-2669
Fax: (852) 865-6843

HK Food Council
(Governing Body of the Hong Kong Food Trades Association)
1/F, CMA Building
64-66 Connaught Rd.
Central, Hong Kong
Tel: (852) 545-6166
Fax: (852) 541-4541
Tlx: 63526 MAFTS HX

HK Fresh Fruits Importers' Association
Rms 401-3, Prosperous Bldg.
48 Des Voeux Rd.
Central, Hong Kong

Tel: (852) 521-1228
Fax: (852) 868-4402

HK Garment Manufacturers' Association
2/F, East Ocean Centre
98 Granville Rd.
Tsimshatsui East, Kowloon
Tel: (852) 367-3392
Fax: (852) 721-7537

HK Hotels Association
508-511 Silvercord Tower Two
30 Canton Rd.
Tsimshatsui, Kowloon
Tel: (852) 369-9577, 375-3838
Fax: (852) 722-7676, 375-7676

HK Information Technology Federation
1/F, Centre Point.
181 Gloucester Rd.
Wanchai, Hong Kong
Tel: (852) 836-3356
Fax: (852) 591-0975

HK Liner Shipping Association
2111 Wing On Centre
111 Connaught Rd.
Central, Hong Kong
Tel: (852) 544-5077
Fax: (852) 815-1350

HK Management Association
14/F, Fairmount House
8 Cotton Tree Drive
Central, Hong Kong
Tel: (852) 526-6516
Fax: (852) 868 4387

HK Office Equipment Association
c/o Kodak (Far East) Ltd.
321 Java Rd.
North Point, Hong Kong

Tel: (852) 564-9333
Fax: (852) 565-7474

**HK Pharmaceutical
Manufacturers' Association**
Cheung Wah Ind Bldg.
Flat A, 12/F
12 Shipyard Lane
Quarry Bay, Hong Kong
Tel: (852) 562-3810, 561-3863
Fax: (852) 563-4018
Tlx: 83085 UNILB HX

HK Plastics Technology Centre
BC717, Hong Kong Polytechnic
Hunghom, Kowloon
Tel: (852) 766-5577
Fax: (852) 766-0131

HK Printers' Association
48-50 Johnston Rd, 1/F
Wanchai, Hong Kong
Tel: (852) 527-1859, 527-5050
Fax: (852) 861-0463

**HK Standards and Testing
Centre**
10 Dai Wang St.
Taipo Industrial Estate
Taipo, New Territories
Tel: (852) 667-0021
Fax: (852) 664-4353

HK Shipowners' Association
12/F, Queen's Centre
58-64 Queen's Rd. East
Wanchai, Hong Kong
Tel: (852) 520-0206
Fax: (852) 865-1582
Tlx: 89157 HKSOA HX

HK Shippers' Council
2707A, Office Tower
Convention Plaza
1 Harbour Rd

Wanchai, Hong Kong
Tel: (852) 824-1228
Fax: (852) 824-0394
Tlx: 73595 CONHK HX

HK Shipping Industry Institute
2607, Alexandra House
16-20 Chater Rd.
Central, Hong Kong
Tel: (852) 526-4294
Fax: (852) 810-6780
Tlx: 85146 SETRA HX

HK Telecom Association
GPO Box 13461, Hong Kong
Tel: (852) 834-3693
Fax: (852) 832-9626

HK Toys Council
4/F, Hankow Centre
5-15 Hankow Rd.
Tsimshatsui, Kowloon
Tel: (852) 723-0818
Fax: (852) 721-3494
Tlx: 30101 FHKI HX

HK Unit Trust Association
12/F, Printing House
No 6 Duddell St.
Central, Hong Kong
Tel: (852) 526-8783
Fax: (852) 868-0224

**HK Watch Manufacturers'
Association**
11/F, Yu Wing Bldg.
64-66 Wellington St.
Central, Hong Kong
Tel: (852) 522-5238
Fax: (852) 810-6614

**Institution of Industrial
Managers**
GPO Box 10221, Hong Kong
Flat 2A, Fu Han Bldg.

No 43-45 Tai Yuen St.
Wanchai, Hong Kong
Tel: (852) 834-8815
Fax: (852) 572-1159

North American Medical Association
1136 Swire House
Chater Rd.
Central, Hong Kong
Tel: (852) 523-2123
Fax: (852) 526-0148

Private Sector Committee on the Environment
1 Queen's Rd.
Central, Hong Kong
Tel: (852) 822-4993, 822-4962
Fax: (852) 845-0113

Real Estate Developers' Association of Hong Kong
1103-04 New World Tower
16-18 Queen's Rd.
Central, Hong Kong
Tel: (852) 522-7245
Fax: (852) 845-2521

Retail Management Association
14/F Fairmont House
8 Cotton Tree Drive

Central, Hong Kong
Tel: (852) 526-6516
Fax: (852) 868-4387
Tlx: 81903 HKMGR HX

Shippers' Association of Hong Kong
2707A, Office Tower
Convention Plaza
1 Harbour Rd.
Hong Kong
Tel: (852) 824-1228
Fax: (852) 824-0394
Tlx: 73595 CONHK HX

Society of Builders of Hong Kong
Rm 801/2, 8/F
On Lok Yuen Bldg.
25 Des Voeux Rd.
Central, Hong Kong
Tel: (852) 523-2081/2
Fax: (852) 845-4749

Textile Council of Hong Kong
Rm 744, Level 7, Star House
3 Salisbury Rd.
Tsimshatsui, Kowloon
Tel: (852) 735-7793
Fax: (852) 735-7795

ANNEX 9

HONG KONG ECONOMIC AND TRADE OFFICES

A good source of information on Hong Kong's tariff regulations, health standards, labelling and packaging requirements are the string of representative offices the Hong Kong government maintains around the world. Unlike HKTDC offices (see Annex 2), these government offices specialize in inter-govermental and regulatory affairs.

GOVERNMENT OFFICES

NORTH AMERICA

New York
Hong Kong Economic and Trade Office
British Consulate General
680 Fifth Avenue
2nd Floor
New York, NY 10019
USA
Tel: (212) 265-8888
Fax: (212) 974-3209

San Francisco
Hong Kong Economic and Trade Office
British Consulate General
222 Kearny Street
Suite 402
San Francisco, CA 94108
USA
Tel: (415) 397-2215
Fax: (415) 421-0646

Washington
Hong Kong Economic and Trade Office

British Embassy
1233 20th Street, N.W.
Suite 504
Washington, DC 20036
USA
Tel: (202) 331-8947
Fax: (202) 331-8958

Toronto
Hong Kong Economic and Trade Office
59th Floor
One First Canadian Place
Toronto, Ontario
Canada M5X 1K2
Tel: (416) 777-2209
Fax: (416) 777-2217

EUROPE

Brussels
Hong Kong Economic and Trade Office
Avenue Louis 228
1050 Brussels
Belgium
Tel: (02) 648-38-33
Fax: (02) 640-66-55

Geneva
Hong Kong Economic and Trade
Office
37-39 rue de Vermont
1211 Geneva 20
Switzerland
Tel: (022) 734-90-40
Fax: (022) 733-99-04

London
Hong Kong Government Office
6 Grafton Street
London W1X 3LB
England
Tel: (071) 499-9821
Fax: (071) 495-5033

ASIA

Tokyo
Hong Kong Economic and Trade
Office
No. 32, Kowa Building
2-32 Minami Azabu
5-chome, Minato-ku
Tokyo, Japan
Tel: 81-3-3446-8099
Fax: 81-3-3446-8126

*INDUSTRIAL PROMOTION
UNITS*

NORTH AMERICA

New York
Industrial Promotion Unit
Hong Kong Economic and Trade
Office
680 Fifth Avenue
22nd Floor
New York, NY 10019
USA
Tel: (212) 265-7232
Fax: (212) 974-3209

San Francisco
Industrial Promotion Unit
Hong Kong Economic and Trade
Office
222 Kearny Street
Suite 402
San Francisco, CA 94108
USA
Tel: (415) 956-4560
Fax: (415) 421-0646

EUROPE

Brussels
Industrial Promotion Unit
Hong Kong Economic and Trade
Office
Avenue Louise 228
1050 Brussels
Belgium
Tel: (02) 648-3966
Fax: (02) 640-6655

London
Industrial Promotion Unit
Hong Kong Government Office
6 Grafton Street
London W1X 3LB
England
Tel: (071) 499-9821
Fax: (071) 495-5033

ASIA

Tokyo
Industrial Promotion Unit
Hong Kong Economic and Trade
Office
No. 32, Kowa Building
2-21 Minami Azabu
5-chome, Minato-ku
Tokyo, Japan
Tel: 81-3-3446-8111
Fax: 81-3-3446-8126

ANNEX 10

SOME FOREIGN TRADE COMMISSIONS IN HONG KONG

AUSTRALIA

Harbour Centre
23-24 floors
25 Harbour Road
Wanchai, Hong Kong
Tel: (852) 573-1881

BRITAIN

Bank of America Tower
9 floor, 12 Harcourt Rd.
Hong Kong
Tel: (852) 523-0176
Fax: (852) 845-2870

CANADA   Federal government

Commission for Canada
13th floor, Tower 1
Exchange Square
8 Connaught Place
Hong Kong
Tel: (852) 847-7414
Fax: (852) 847-7441

Canadian Embassy
South China Trade Office
13th floor, Tower 1
8 Connaught Place
Hong Kong
Tel: (852) 847-7478
Fax: (852) 847-7436

Provincial governments   Alberta
Room 1003-4, Admiralty
Centre, Tower 2
18 Harcourt Road
Hong Kong
Tel: (852) 528-4729
Fax: (852) 529-8115

British Columbia
901 Hutchison House
10 Harcourt Road
Hong Kong
Tel: (852) 845-1155
Fax: (852) 845-4114

Manitoba
902A China Building
29 Queen's Road, Central
Hong Kong
Tel: (852) 523-3375
Fax: (852) 845-3076

Ontario
Rooms 906-908
Hutchison House
10 Harcourt Road
Hong Kong
Tel: (852) 845-3388
Fax: (852) 845-5166

	Québec Bond Centre, East Tower 19/F, 89 Queensway Hong Kong Tel: (852) 810-9332 Fax: (852) 845-3889
FRANCE	Admiralty Centre Tower II, 26/F 18 Harcourt Road Hong Kong Tel: (852) 529-4316 Fax: (852) 861-0019
GERMANY	21/F, United Centre 95 Queensway Hong Kong Tel: (852) 529-8855 Fax: (852) 865-2033
ITALY	Room 805 Hutchison House 10 Harcourt Road Hong Kong Tel: (852) 522-0033 Fax: (852) 845-9678
NEW ZEALAND	3414, Jardine House Connaught Road Hong Kong Tel: (852) 525-5044 Fax: (852) 845-2915
UNITED STATES OF AMERICA	26 Garden Road Hong Kong Tel: (852) 523-9011 Fax: (852) 845-9800

ANNEX 11

CANADIAN REGIONAL INTERNATIONAL TRADE CENTERS

Newfoundland

215 Water St., Suite 504
P.O. Box 8950
St. John's, Newfoundland
A1B 3R9
Tel: (709) 772-5511
Fax: (709) 772-2373

Prince Edward Island

Confederation Court Mall
134 Kent Street, Suite 400
P.O. Box 1115
Charlottetown, P.E.I.
C1A 7M8
Tel: (902) 566-7400
Fax: (902) 566-7450

Nova Scotia

Central Guarantee Trust Building
1801 Hollis Street
P.O. Box 940, Station M
Halifax, Nova Scotia
B3J 2V9
Tel: (902) 426-7540
Fax: (902) 426-2624

New Brunswick

Assumption Place
770 Main Street
P.O. Box 1210
Moncton, New Brunswick
E1C 8P9
Tel: (506) 851-6452
Fax: (506) 851-6429

Quebec   Stock Exchange Tower, Suite 3800
                            800 Victoria Square
                            P.O. Box 247
                            Montreal, Quebec
                            H4Z 1E8
                            Tel: (514) 283-8185
                            Fax: (514) 283-8794

Ontario   Dominion Public Building, 4th Floor
                            1 Front Street West
                            Toronto, Ontario
                            M5J 1A4
                            Tel: (416) 973-5053
                            Fax: (416) 973-8161

Manitoba   8th Floor
                            330 Portage Avenue
                            P.O. Box 981
                            Winnipeg, Manitoba
                            R3C 2V2
                            Tel: (204) 983-8036
                            Fax: (204) 983-2187

Saskatchewan   119-4th Avenue South, Suite 401
                            Saskatoon, Saskatchewan
                            S7K 5X2
                            Tel: (306) 975-5315
                            Fax: (306) 975-5334

                            1955 Smith Street, 4th Floor
                            Regina, Saskatchewan
                            S4P 2N8
                            Tel: (306) 780-5020
                            Fax: (306) 780-6679

Alberta   Canada Place, Suite 540
                                   9700 Jasper Avenue
                                   Edmonton, Alberta
                                   T5J 4C3
                                   Tel: (403) 495-2944
                                   Fax: (403) 495-4507

                                   International Trade Centre
                                   510-5th Street S.W., 11th Floor
                                   Calgary, Alberta
                                   T2P 3S2
                                   Tel: (403) 292-6660
                                   Fax: (403) 292-4578

British Columbia   International Trade Centre
                                   Scotia Tower, Suite 900
                                   650 West Georgia Street
                                   P.O. Box 11610
                                   Vancouver, British Columbia
                                   V6B 5H8
                                   Tel: (604) 666-0434
                                   Fax: (604) 666-8330

Yukon   Industry, Science and Technology Canada
                                   (ISTC)
                                   108 Lambert Street, Suite 301
                                   Whitehorse, Yukon
                                   Y1A 1Z2
                                   Tel: (403) 668-4655
                                   Fax: (403) 668-5003

Northwest Territories   Industry, Science and Technology Canada
                                   (ISTC)
                                   Precambrian Building, 10th Floor
                                   4922-52nd Street, P.O. Box 6100
                                   Yellowknife, Northwest Territories
                                   X1A 2R3
                                   Tel: (403) 920-8568
                                   Fax: (403) 873-6228

ANNEX 12

ASIA PACIFIC FOUNDATION OF CANADA OFFICES

HEAD OFFICE
666-999 Canada Place
Vancouver, B.C.
Canada V6C 3E1
Tel: (604) 684-5986
Fax: (604) 681-1370

REGIONAL OFFICES

QUEBEC
525 Cherrier Street East
Montreal, Quebec
H2L 1H2
Tel: (514) 982-9300
Fax: (514) 982-9060

ONTARIO
65 Queen St. West
Suite 1100
Toronto, Ontario
Canada, M5H 2M5
Tel: (416) 869-0541
Fax: (416) 869-1696

SASKATCHEWAN
Hong Kong Bank Building
1874 Scarth Street
Regina, Saskatchewan
Canada, S4P 4B3
Tel: (306) 791-8778
Fax: (306) 359-7066

ALBERTA   40 McDougall Centre
                        455 6th Street S.W.
                        Calgary, Alberta
                        Canada, T2P 4E8

ASIAN OFFICES

SINGAPORE   80 Anson Road #15-02
                        IBM Towers
                        Singapore 0207
                        Tel: (65) 225-7346
                        Fax: (65) 222-7439

TAIWAN   13/F, 365 Fu Hsing North Road
                        Taipei, 10483 Taiwan
                        Tal: (02) 713-7268
                        Fax: (02) 712-7244

JAPAN   3/F, Place Canada
                        3-37 Akasaka 7-chome
                        Minato-ku, Tokyo 107, Japan
                        Tel: (03) 5410-3838
                        Fax: (03) 5410-3020

INDEX

INDEX

Titles soon to be released in the
EXPORTER'S DISTRIBUTION PRIMER
series.

- JAPAN
- THAILAND/INDOCHINA

Each book covers:

- the political, social and economic trends which directly affect distribution in the booming and increasingly integrated markets,

- the purchasing behaviour of importers of capital goods and consumer products,

- the marketing strategies used by distributors,

- a step-by-step approach to effectively selecting a distributor,

- crucial contacts : trade associations, credit investigation firms, government trade offices, law and accounting firms, market research organizations.

Canada	Groupe
Communication	Communication
Group	Canada
Publishing	Édition

Fax this form to (819) 956-5539 or mail it to:
Publications Management,
Canada Communication Group-Publishing,
45 Sacré-Coeur Blvd, Room A2404, Hull, Quebec,
CANADA K1A 0S9

- ✂

YES, I would like to be informed when the new titles in
The Exporter's Distribution Primer series become available.

Name: _____

Organization:_____

Address: _____

_____

City: ————————————————————————

Province: _____ Postal Code: _____

Country: _____

This is not an order form.

ALSO AVAILABLE...

THE TAIWAN BUSINESS PRIMER

The primer covers:
- guidance on local customs and how to develop successful business relationships,
- clear summaries of Taiwan's political, social and economic trends,
- profiles of major industries and professions,
- valuable information on Taiwanese business practices:

business values, the legal framework, employment, distribution, exporting, sourcing and investment,
- useful addresses: trade associations, credit investigation firms, audit corporations, law and patent firms,
- advice on getting around.

This primer is the result of a project sponsored by the *Asia Pacific Foundation of Canada.*

$20	(**US$26** - Other countries)	12,5 x 18 cm
Catalogue No. K49-1-1991E	Paperbound	314 pages

Également disponible en français sous le titre *TAIWAN LE GUIDE DES GENS D'AFFAIRES* (N° de catalogue K49-1-1991F). — — — — — — — — — — — — — — — ✂

ORDER FORM

YES, send me

___ copies of **HONG KONG/SOUTH CHINA: The Exporter's Distribution Primer** (Cat. No. K49-2-1992E) at the unit cost of $35 (US$45.50 for other countries)

___ copies of **THE TAIWAN BUSINESS PRIMER** (Cat. No. K49-1-1991E) at the unit cost of $20 (US$26 for other countries).

Add to this amount the shipping and handling fees (in Canada, add 7% GST).

Please print

Name : _____

Firm: _____

Address: _____

City: _____

Province : _____ Postal Code: _____

Country: _____ Telephone : ( )

Refer to the following pages for *Method of Payment*,
Shipping and Handling Fees and *Distributors' Addresses*.

FOR ORDERS FROM CANADA

end your orders to:
**Canada Communication
Group - Publishing
Ottawa CANADA
K1A 0S9**

Tel.: 819) 956-4802
Fax: (819) 994-1498

*These publications are also available
at or may be ordered through
Canadian booksellers.*

METHOD OF PAYMENT

Purchase Orders are accepted from governments, registered companies
and educational institutions. Federal government must provide official
purchase orders. All others must prepay.

Purchase Order No.: [   ]

◯ Cheque/Money Order (payable to the Receiver
General for Canada) enclosed
◯ **Visa/MasterCard**

Account No. [   ]

Expiry Date: _____

Signature : _____

Shipping and Handling Fees

Order Value	Fees
$5.01 to $25	$3.50
$25.01 to $75	$5.40
$75.01to $200	$10.50
Over $200	6% of the total order value

FOR ORDERS FROM THE UNITED STATES

Send your orders to:

**International Specialized
Book Services**
5602 NE Hassalo Street N.E.
Portland OR 97213

Tel.: 1-800-547-7734 (toll free)
Tel.: (503) 287-3093 (in Oregon)
Fax : 503-284-8859

**Accents Publications
Services Inc.**
911 Silver Spring Avenue,
Suite 202
Silver Spring, MD 20910

Tel.: 301-588-5496
Fax : 301-588-5249

FOR ORDERS FROM OTHER COUNTRIES

Send your orders to:

Canadian Books Express
The Abbey Bookshop
29, rue de la Parcheminerie
75005 Paris
FRANCE

Tel.: 46.33.16.24
Fax : 46.33.03.33

Asia Book House
16/17 Bangla Bazar
Dhaka 1100
BANGLADESH

Tel : 91-800-2-245650
Fax : 91-800-2-833983

Books Express
P.O. Box 10
Saffron Walden
Essex CB11 4EW
UNITED KINGDOM

Tel.: 0044-799-513726
Fax : 0044-799-513248

Academic Book Store
P.O. Box 128
00100 Helsinki
FINLAND

Tel : 358.0.121.4325
Fax : 358.0.121.4441